ON THE ORIGINALITY OF SPECIES

The Convergence of Evolutionary Science and Baha'i Teachings

BRYAN DONALDSON

Akka Publishing House

ISBN 979-8-3935-8010-0

First edition 2023

Published in the United States by
Akka Publishing House
Portland, OR

originalityofspecies@gmail.com

Look at the world and ponder a while upon it. It unveileth the book of its own self before thine eyes and revealeth that which the Pen of thy Lord, the Fashioner, the All-Informed, hath inscribed therein. It will acquaint thee with that which is within it and upon it and will give thee such clear explanations as to make thee independent of every eloquent expounder.

—Baha'u'llah

TABLE OF CONTENTS

PREFACE

This is not the book I set out to write. This is the book I assumed had already been written and sought to read. In fact, I started 12 years ago by writing a blog article that grew too large and was never posted. I write as a hobby and tend to gravitate toward perplexing issues. The topic of evolution was the last of a series of thorny questions involving the Baha'i Faith that I wanted to write about, so in true procrastinating style, I anticipated the immense work ahead and avoided it. Eventually I read several books and articles by other Baha'is on the subject in addition to sifting through the ocean of texts on biology.

My investigation took on three parts: a thorough review of the writings of 'Abdu'l-Baha and Shoghi Effendi, a review of the extensive discourse on evolution by Baha'i authors, and an exploration of new and theoretical shifts in science that could find agreement with a unique line of descent for humans. Two Baha'i authors have commented on the need for such a review in light of 'Abdu'l-Baha's comments, but the task until now has been absent.

I present the results without any authority in the fields of science or religion. Of course, any interpretation of the Baha'i Writings reflects my personal understanding and not the Baha'i position. This book is part of "an exchange of views conducted in a consultative spirit"* for anyone curious about 'Abdu'l-Baha's writings on human origin.

---

* Universal House of Justice (21 February 2016)

Just as scientific research advanced over the years of writing, so did the understanding among Baha'is. The Universal House of Justice wrote a letter in 2016 addressing how the principle of the harmony of science and religion is applied to the topic of evolution, the first of its kind. This, and the 2014 retranslation of *Some Answered Questions* required major revisions to how I address the Baha'i perspective. A response from the Universal House of Justice about this book (See Appendix I) and feedback from reviewers instigated another major revision period in 2019.

Some basic knowledge of both biology and the Baha'i Faith is assumed for any reader, even though both subjects are approached gradually and avoiding jargon (and a glossary is provided). This limitation to the audience touches on the two extremes I was trying to avoid.

The first extreme is the Baha'i who would declare that my book vindicates the Baha'i religion. Proposing a biological basis for independent descent avoids a potentially catastrophic conclusion about 'Abdu'l-Baha making false statements and avoids awkward attempts at reinterpreting the statements to fit science.

The second extreme is the science-minded individual who would oppose the existence of such a book and not bother reading a page. Once clear that I am examining scientific ideas in the context of a religious doctrine, they would assume a lack of objectivity and logic.

Both readers may agree or disagree without evaluating the data I present. I urge everyone to read and investigate with a sceptical and open mind. There are no appeals to supernatural events or intelligent design, and I present ideas that are either already supported by established publications or need to be confirmed scientifically. Such a sceptic gave me this feedback: "As the piece moves along, it becomes clear that you did your homework."

Special thanks to my wife Jenna for putting up with me; to my daughter Nahla who allowed me to write while she slept on my lap as a baby, then grew up and sketched two pieces of artwork used in the book; to Keven Brown for offering advice; to Skye Lininger, Alex Hanson, and Alyssa Harris for providing superb and free editing; and to Roger Neyman and others who disagreed with me but cared enough to give feedback.

Bryan Donaldson 2023

CHAPTER 1

EVOLUTION AND THE BAHA'I FAITH

Science and religion

Science deals with understanding the physical world and religion advances the social one. These two systems of knowledge are mostly independent of each other, but of course they occasionally overlap and create much debate. Such tension is not easily disentangled, and it is most pronounced on the topic of how and why humans came about.

In public talks elaborating the basic principles of the Baha'i Faith, 'Abdu'l-Baha described four methods of comprehension.[1] That is, there are four reasons why someone believes a thing to be true. For example, what can be known about the essence of a beet?

Observation. I grow beets in my garden. I know beets are real because I can see them, touch them, smell them, and taste them. If a beet drops, I can hear it. Simple observation is the most basic way of knowing a thing is true, but it can still produce false results. A mirage may appear, the earth seems to be still, a whirling item appears as a circle, and a schizophrenic brain will observe things that are not real.

Reason. Beets taste good to me. My sister likes food that I like. Therefore, beets taste good to my sister. Logic and reasoning allow for a new thing to be proven or disproven based on known things. However, this method can produce false results. Any logical statement that is absolutely true doesn't say much of anything useful (beets taste good to me) and the new thing that is known comes from making reasonable assumptions (my sister likes food that I like). Ancient philosophers often came to opposing conclusions or refuted arguments that they had previously upheld. The variability shows that it is not a perfect method.

Tradition. A good way to eat beets is chopped and roasted with olive oil. I know this because my family made them that way when I was growing up. I have never found a better way to eat beets and I do not feel the need to spend time trying new recipes. Traditions that are passed down from generation to generation contain wisdom and truth. Similarly, scriptures advance virtue and guide civilizations. However, this method can produce false results. Erroneous traditions may be carried on without being questioned, the text of a holy book can be interpreted in many ways, or the true meaning of traditions may be lost with changing social context.

Inspiration. Beets are beautiful, they look good in my salad, and they're healthy! The beet is an archetype, a stable pattern that is observed in the many examples of living beets. I know these things not from observation, logic, or tradition. This kind of knowledge is established through intuition and the promptings of the heart. The comprehension gained through inspiration creates more certainty than the first three methods but is liable to error. Animalistic or selfish promptings might also appear to be intuition. Inspiration alone cannot distinguish truth from falsehood.

Perceived truths coming from these four methods are sometimes in conflict, prompting the need to investigate the source of error. When some new idea appears to conflict with established knowledge, some may fall into cognitive dissonance instead of allowing their assumptions to be challenged. Others may review the observations, logic, and traditions while meditating on the issue to reveal a new reality or reinforce the established one. Turning to consultation can also help. What conclusion has a trusted friend come to? What is the consensus?

This investigation of truth applies to both scientific pursuits and religious beliefs. How many adherents follow the religion of their parents without investigating other views, or follow harmful practices out of a duty to tradition? How many scientists accept uncritically what they have been taught, or sacrifice objectivity for personal gain?

A critic might suggest that religion claims to usurp the authority of science, so that religion always wins in a conflict between the two. In the Baha'i Faith, this is not the case. 'Abdu'l-Baha condemned belief in religious doctrines that are opposed to science and reason,

> If religious belief and doctrine is at variance with reason, it proceeds from the limited mind of man and not from God; therefore, it is unworthy of belief and not deserving of attention; the heart finds no rest in it, and real faith is impossible.[2]

Again,

> Every religion which is not in accordance with
> established science is superstition. Religion must
> be reasonable. If it does not square with reason,
> it is superstition and without foundation.[3]

And again,

> A question or principle which is religious in its
> nature must be sanctioned by science. Science must
> declare it to be valid, and reason must confirm it
> in order that it may inspire confidence. If religious
> teaching, however, be at variance with science
> and reason, it is unquestionably superstition.[4]

And once again,

> Put all your beliefs into harmony with science;
> there can be no opposition, for truth is one.[5]

In a letter on behalf of Shoghi Effendi to an individual asking about cancer treatments, we find a definitive statement about the Baha'i relationship to science.

> But as we are a religion and not qualified to pass on
> scientific matters we cannot sponsor different treatments.[6]

Religion and science both benefit mankind and must be in harmony. That is not to say that someone wearing a 'scientist' badge can declare something to be true and expect Baha'is to accept it. Conflicting perceptions of reality must find reconciliation, and the reconciliation may involve scripture or our understanding of scripture being wrong, but reconciliation may also involve reinvestigating the conclusions of scientists.

Just as there is a distinction between divine revelation and the many interpretations of it by believers, there is also a distinction between scientific fact and the many ideas and theories of scientists. Science is not an isolated, impartial entity that exists outside of its social context. Therefore, its understanding of truth is contingent and evolving. Religion cannot uncritically agree with all scientific pronouncements at all times for the simple reason that ideas and theories of scientists are sometimes contradictory and subject to continual change to better explain new discoveries.

'Abdu'l-Baha was once asked by a believer in Paris about how to harmonize scientific theories with the ideas of religion. He responded that theories coming from "sound scientific thinking" are in agreement with religion, and "if a scientific theory does not correspond with the divine verses, it is certain that it is the essence of error."[7] He gave the example of certain Qur'anic verses appearing to indicate that the sun is fixed and the planets are in motion, which, of course, contradicted common understanding when the Qur'an was written and were proven correct centuries later. 'Abdu'l-Baha also gave the example of erroneous religious beliefs with "no basis in fact", such as the idea that the soul dwells on other planets.[8]

Evolution

Among adults in the United States in 2015, only about one third believed that humans evolved by natural processes. The remainder were either unsure (5%), did not believe that humans evolved at all (31%), or believed that evolution was divinely guided (24%).

The disbelief in natural evolution was closely tied to religious belief. Among the religiously unaffiliated, 67% believed in evolution by natural selection, while that number dropped to 12% among white evangelical Christians.[9]

It is only natural to inquire about the Baha'i position on the issue, with an eye for any deviance whatsoever from standard evolutionary theory. Let's see how a Baha'i might answer some standard probing questions.

- Have humans existed in their present form since their first appearance on earth?
- Was the story of Adam and Eve an account of the first humans?
- Are human beings the creation of natural laws?
- Do species evolve by natural selection?
- Did humans develop over many millions of years from a primitive form?
- Do modern humans and chimpanzees share a common ancestor about 8 million years ago?

As a Baha'i, my answers are:

- No
- No
- Yes
- Yes
- Yes
- It's complicated

The complication can be summed up in this way: 'Abdu'l-Baha, one of three central figures of the Baha'i Faith, made statements about human origin that at least superficially appear to conflict with standard evolutionary theory. He described humans as evolving from a primitive form, but in a way that humans were never animals, leading many to interpret the statements as indicating a kind of parallel evolution. Many Baha'is have viewed it as an issue that will be validated by new scientific discoveries. However, 'Abdu'l-Baha also said that religious belief conflicting with science should be viewed as superstition and disregarded, leaving some ambiguity in how Baha'is understand human evolution. This ambiguity leaves the field open for discourse on how Baha'is might reconcile the apparent contradiction between these statements, the harmony of religion and science, and the prevailing theory of universal common ancestry.

Modern Baha'i authors have tried to reconcile these ideas in many ways and the clear majority consider established science as fact regarding the common ancestry of humans and chimpanzees. Explanations range from interpreting 'Abdu'l-Baha's statements in a way that is consistent with common ancestry, to suggesting that 'Abdu'l-Baha simply made inaccurate statements. The most common understanding is nicely summarized by Baha'i author Gary Matthews, who wrote, "the apparent contradiction is nothing more than a question of semantics: perhaps 'Abdu'l-Baha is merely dating man's beginning as a distinct species from the soul's first appearance, to emphasize that we do not derive our higher spiritual nature from our animal forebears".[10]

This book takes an untrodden approach by exploring the current science of evolution in the context of 'Abdu'l-Baha's comments. Although it deals with materialistic mechanisms of evolution, it also explores the spiritual nature of human reality. Ultimately, the details of human ancestry are not nearly as important as human virtue and its place as the crowning achievement of an evolutionary process.

The apparent conflict over human origin is rooted in statements by 'Abdu'l-Baha and the social context of early 20th-century Darwinism. A brief review of both will be a crucial bedrock for any exploration of the tension.

The Baha'i Faith

The Baha'i Faith was founded in 1863 by Baha'u'llah (1817–1892) and preceded by a religious movement in Persia founded by the Bab (1819–1850). The Bab's core message was that humanity stood at the threshold of a new era, and that soon a great messenger of God would appear. Several years after the Bab and thousands of his followers were executed by the Persian authorities, Baha'u'llah, exiled in Iraq, announced that he was the figure anticipated by the Bab, making claim to a revelation from God in fulfilment of the messianic expectations of Islam, Christianity, and other past religions.

In addition to peace, love, and unity, Baha'u'llah's teachings also encourage the independent search for truth, the unity of all religions, the equality of men and women, the elimination of extremes of wealth and poverty, the harmony that must exist between religion and science, and the fundamental unity of the human race. While admonishing certain world leaders of the time who were squandering lives in fruitless wars, he taught the necessity of a world tribunal to resolve disputes between nations and encouraged the adoption of a universal auxiliary language, a uniform system of weights and measures, and a compulsory education for all children.

From the time of his declaration, Baha'u'llah spent his entire life as a religious prisoner of the Ottoman Empire. Upon his death, Baha'u'llah passed authority on to his son 'Abdu'l-Baha (1844–

1921), who was eventually released from prison and allowed to travel, spreading his father's teachings. The Bab, Baha'u'llah, and 'Abdu'l-Baha are the three central figures of the Baha'i Faith and their authenticated writings are sources of scripture. Baha'u'llah's great-grandson Shoghi Effendi later wrote that 'Abdu'l-Baha was "the perfect Exemplar" of Baha'u'llah's Faith and "endowed with super-human knowledge".[11]

After 'Abdu'l-Baha's own passing, authority was left with both the elected governing body referred to by Baha'u'llah as the Universal House of Justice, and the hereditary Guardianship. The appointed Guardian and the Universal House of Justice are both considered infallible within their respective broad spheres of jurisdiction, the Guardian providing authoritative interpretation, and the Universal House of Justice providing flexibility and adjudication on questions that are not expressly discussed in the writings of the central figures.

'Abdu'l-Baha appointed his grandson, Shoghi Effendi, as the first Guardian. Although he allowed for further Guardians to be appointed, there were no eligible appointees at the time of Shoghi Effendi's death. He will remain both the first and last holder of the position and the Universal House of Justice is the highest governing institution of the international Baha'i community. It formed in 1963 and is elected every five years by members of all National Spiritual Assemblies of the world, who are themselves elected by a system of delegates from Baha'i communities throughout each nation.

On Darwin

Before Charles Darwin, the majority of naturalists viewed God's special creation as the only explanation for the complexity of life on earth. They almost universally accepted that species were static and timeless, beginning at the time of creation and remaining in basically the same form, with only minor adaptations. Darwin and a handful of others boldly moved the academic community in a new direction and their work laid the foundation for one of the greatest revolutions in science. Social applications of Darwin's research by others were equally profound. The early 20th century witnessed attempts to apply principles of natural selection to human populations, in a kind of social evolution where weaker peoples would be eliminated by stronger ones (e.g. eugenics, extreme capitalism).

Darwin observed that a population undergoes minor biological changes from generation to generation and those traits that give some advantage in survival or reproduction get passed on to future generations. Then he proposed a theory to explain speciation – the process by which new species arise. He posited that all the various species of life on earth arose from the branching of previous generations by means of natural selection. A variant of this idea is to say that all life on earth arose from a single universal common ancestor, or as Darwin himself put it, "probably all of the organic beings which have ever lived on this earth have descended from some one primordial form…"[12]

'Abdu'l-Baha and evolution

In 1912, 'Abdu'l-Baha made a historic trip to Europe and America, spreading his father's teachings, while giving talks to religious congregations, social movements, and other gatherings. During several of the talks, 'Abdu'l-Baha addressed the debate over human origins. The trip took place while Charles Darwin's theory of evolution was gradually gaining acceptance in the minds of scientists, but the trip was still 20 years before the theory was widely accepted among a variety of disciplines and 40 years before the discovery of DNA.

'Abdu'l-Baha taught that "the world of existence – that is, this endless universe – has no beginning",[13] that the story of the six days of creation "is not to be taken literally",[14] that the days of creation "represent time spans of millions of years",[15] and that the story of Adam and Eve is full of symbology and not to be taken as an historical account.[16] He does not dispute the gradual development of species by natural selection, but 'Abdu'l-Baha does clash with part of Darwin's theory that humans evolved from animals (see later this chapter).

As the theory of evolution grew in prominence, so did the tension with 'Abdu'l-Baha's statements. The Baha'i principle of the harmony of science and religion required a resolution. In an unrelated but similar example, 'Abdu'l-Baha used the term 'ether' as the medium through which light propagates, but the idea of ether was later replaced by other concepts and the term fell out of favor with scientists. 'Abdu'l-Baha's use of the term creates a kind of tension, but on closer analysis he was using language as understood at the time to make unrelated points, and he described ether as an intellectual reality with no physical properties, so there has never been a serious conflict over the idea.[17]

To investigate the comments on evolution and reconcile the tension with science, there are a few questions to address: Is the text authentic? Is the text properly translated? What was the social context in which it was expressed? Can the text be interpreted in a way that is consistent with science? Would an incorrect scientific statement bring into question the divine origin of the entire religion? Is the contemporary scientific consensus correct?

Authority of sources

Islam has an entire category of literature, called *hadith*, representing thousands of personal accounts of Muhammad's life. Highly authentic hadiths are still not as authoritative as the Qur'an itself, but the body of literature altogether adds a rich source of teachings and stories to the religion.

Similarly, there are varying degrees of authenticity of 'Abdu'l-Baha's statements. The Baha'i Faith has a category of literature that is not as reliable or authoritative as actual religious texts, but it adds to an understanding of the religion. Many people met with 'Abdu'l-Baha and returned with stories of their encounter, even publishing memoirs. Shoghi Effendi later commented on the authority of these kinds of sources:

> ... the notes of the pilgrims should be for their own personal use and bear absolutely no authority.[18]

'Abdu'l-Baha also travelled throughout Europe and America giving public talks. These talks were delivered in Persian while a linguist conveyed the English version. Sometimes only the English translation was recorded, sometimes the original Persian was recorded, and sometimes the original Persian was reviewed and approved by 'Abdu'l-Baha.* It is only this last category that may be called fully authenticated and part of scripture.

Shoghi Effendi also elaborated on the varying authority of these recorded talks in a letter written on his behalf in 1931,

> Those talks of ['Abdu'l-Baha] that were later reviewed by Him, corrected or in some other form considered authentic by Himself... these could be considered as Tablets and therefore be given the necessary binding power. All the other talks... do not fall under this category and could be considered only as interesting material to be taken for what they are worth.[19]

Unlike Islamic hadiths, whose ambiguous authority brought a degree of chaos to interpreting scripture, the Baha'i system of authentication is aided by an International Archives and a Research Department under the Universal House of Justice that can identify original texts.

Someone wishing to resolve the apparent discord between Baha'i texts and Darwinian evolution should first establish the degree of authenticity of the sources. Finding that the sources are pilgrims' notes or "mere talks" that lack authentication would notably dampen the appearance of conflict.

Most of the explicit references to evolution come from *Some Answered Questions*, which is a compilation of table talks that were

---

* The introduction to the 1922 Edition of *Promulgation of Universal Peace* wrote, "This treasury of His words is a compilation of informal talks and extemporary discourses delivered in Persian and Arabic, interpreted by proficient linguists who accompanied Him, and taken stenographically in both Oriental and Occidental tongue."

recorded by an American Baha'i while in Haifa. The only other significant source on the subject comes from *The Promulgation of Universal Peace*, which is a compilation of talks given by 'Abdu'l Dalia as he travelled across America. The subjects of biological origin and Darwinian theory do not appear explicitly elsewhere in the available writings of the central figures in English.

Promulgation of Universal Peace

On the subject of talks recorded in *The Promulgation of Universal Peace*, the argument for dismissing their authority may or may not be valid. The Secretariat of the Universal House of Justice conveyed in 1987 that,

> For many of His addresses included in "*The Promulgation of Universal Peace*" and "*Paris Talks*"... no original authenticated text has yet been found... In the future each talk will have to be identified and those which are unauthenticated will have to be clearly distinguished from those which form a part of Baha'i Scripture. This does not mean that the unauthenticated talks will have to cease to be used – merely that the degree of authenticity of every document will have to be known and understood.[20]

One such talk of great interest to the topic of human evolution was delivered at the Open Forum in San Francisco, 10 October 1912. What is the degree of its authenticity? Unfortunately, the answer is not conclusive. The English translation of the talk is recorded in *Promulgation of Universal Peace*, but all such translated talks are imprecise words of an interpreter in front of an audience, not a carefully rendered translation of the original Persian. The talk at the Open Forum has the Persian transcription saved for posterity, but it has not been officially retranslated, and no comprehensive review of the authority of those talks has been performed.[21] Of course, having a verbatim record of what was said in Persian is very reliable,[22] but to be considered authenticated and part of scripture, it must have received the sanction of 'Abdu'l-Baha.

Did this talk receive such a sanction? Maybe. Mahmud-i-Zarqani, the chronicler of 'Abdu'l-Baha's travels in America, wrote that,

> At every gathering, whether for Baha'is or non-Baha'is, several stenographers, as well as the Persian secretaries, were in attendance. The English translations were published soon after the address itself but the Persian originals taken down by us verbatim had to be submitted to 'Abdu'l-Baha for correction. Because of His heavy schedule, He had little time for this, so the originals were often delayed in their publication.[23]

The Persian transcription of the talk at the Open Forum was published, and was provisionally retranslated by Keven Brown in *Evolution and Baha'i Belief* (see Appendix C). The revised translation corrects some inaccuracies from the English translation (e.g. "in the protoplasm, man is man" was not in the original), but the tension clearly remains. Whether the Persian transcript was ever approved by 'Abdu'l-Baha remains unconfirmed but likely.

Some Answered Questions

Regarding *Some Answered Questions*, the richest source of statements on human origin, the same letter on behalf of the Universal House of Justice describing the authenticity of sources mentioned,

The original of *Some Answered Questions* in Persian
is preserved in the Holy Land; its text was read
in full and corrected by Abdu'l-Baha Himself.

Shoghi Effendi mentions it as a source of talks that were,

reviewed by Him, corrected or in some other
form considered authentic by Himself . . . these
could be considered as Tablets and therefore
be given the necessary binding power.[24]

Moreover, *Some Answered Questions* is not merely an authenticated text, but it sits prominently among the core scripture of the religion. 'Abdu'l-Baha encouraged others to read the book and to translate it into many languages. In letters written on his behalf, Shoghi Effendi mentions *Some Answered Questions* as one of several books that every Baha'i should "master", "read over and over again", and "be able to explain their contents to others".[25]

The talks of 'Abdu'l-Baha

Anyone taking this subject seriously must thoroughly study the original texts. No currently living individual has the authority to forge new Baha'i doctrine or make official interpretations.

The subject of human evolution is discussed at length and directly in *Some Answered Questions*, nos. 46–51, and *The Promulgation of Universal Peace* (2012 ed.), pp. 504-508.

For a more thorough review of less-direct references that would aid the study, read *Some Answered Questions* no. 64, *The Promulgation of Universal Peace* pp. 78-79, 86-87, 92-93, 109-113,

315-316, 425-427, 492-502, *Paris Talks,* nos. 2, 23, 28, 29, 31, 54, and the *Tablet to Dr. Forel.*

Some Answered Questions is a collection of table talks recorded by an American visitor to Palestine in 1904–06. In question number 46, 'Abdu'l-Baha said,

> We now come to the question of... whether man has come from the animal kingdom.
>
> This idea has entrenched itself in the minds of certain European philosophers, and it is very difficult now to make its falsity understood; but in the future it will become clear and evident, and the European philosophers will themselves recognize it. For in reality it is an evident error.

In number 47, 'Abdu'l-Baha said,

> ... from the beginning of man's existence on this planet until he assumed his present shape, form, and condition, a long time must have elapsed, and he must have traversed many stages before reaching his present condition. But from the beginning of his existence man has been a distinct species.

As recorded in *Some Answered Questions,* number 49, he was asked, "What do you say regarding the theory of the evolution of beings to which certain European philosophers subscribe?" To which he responded, "this question comes down to the originality or non-originality of the species, that is, whether the essence of the human species was fixed from the very origin or whether it subsequently came from the animals. Certain European philosophers hold that species evolve and can even change and transform into other species." He then summarizes the proofs prevalent at the time for humans evolving from animals, separating the observable facts from the conclusions of those philosophers. The observations were, he says, that the geological record shows that plants preceded animals, which preceded humans; that the geological record shows species improving and becoming hardier; and that humans and other creatures have vestigial organs. It is clear that he disagrees with the conclusion from this data that species can transform into other species over time. He lists three answers against such an idea:

The first answer to this argument is that the antecedence of animals to man is not a proof that the essence of the human species was altered or transformed or that man came from the animal kingdom. For so long as it is acknowledged that these different beings have appeared in time, it is possible that man simply came into existence after the animal. Thus we observe in the vegetable kingdom that the fruits of different trees do not appear all at once; on the contrary, some appear earlier in the season and others later. This priority is not a proof that the later fruit of one tree was produced from the earlier fruit of another.

Secondly, these minor traces and vestigial limbs might have some great underlying wisdom which the human mind has so far been unable to fathom. How many things are found in this world whose underlying wisdom to this day has not been grasped! Thus, it is said in physiology—the science of the relations of the body's organs—that the underlying wisdom and cause of the differences in the colouration of animals and of human hair, or of the redness of the lips, or of the variety of the colours of birds, are still unknown and remain hidden and concealed. But it has been discovered that the blackness of the pupil of the eye is due to its absorbing the rays of the sun, for if it were of another colour—say, uniformly white—it would not absorb these rays. Now, so long as the wisdom underlying the things that we have mentioned is unknown, one may well imagine that the reason and wisdom of the vestigial limbs, whether in the animal or in man, is also unknown. Such an underlying wisdom of course exists, even though it may not be known.

Thirdly, even if we were to suppose that certain animals, or even man, once possessed limbs which have now disappeared, this would not be a sufficient proof of the transformation of the species. For man, from the conception of the embryo until the attainment of maturity, assumes different forms and appearances. His appearance, form, features, and colour change; that is, he passes from form to form and from appearance to appearance. Yet, from the formation of the embryo he belongs to the human species; that is, it is the embryo of a man and not of an animal. But at first this fact is not apparent; only later does it become plain and visible.

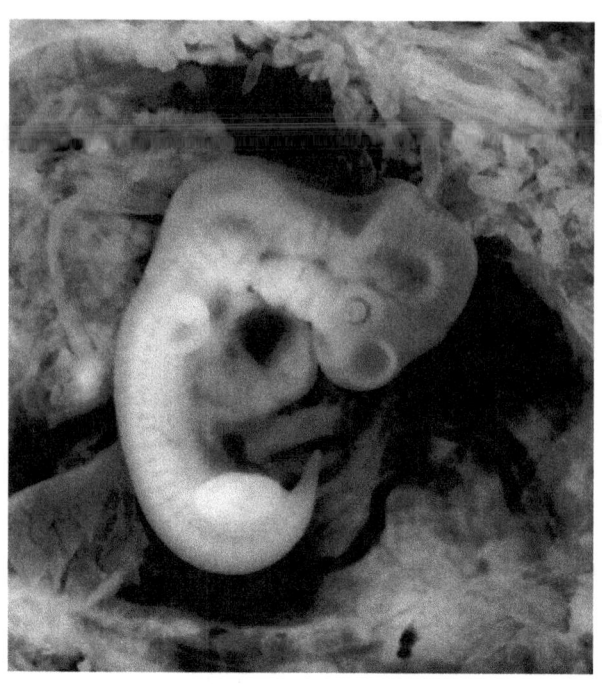

Human embryo at 7 weeks of pregnancy. Source: Wikimedia.

For example, let us suppose that man once bore a
resemblance to the animal and that he has since evolved
and transformed. Accepting this statement does not
prove the transformation of species, but could instead
be likened to the changes and transformations that
the human embryo undergoes before reaching its full
development and maturity, as was earlier mentioned. To
be more explicit, let us suppose that man once walked
on all fours or had a tail: This change and transformation
is similar to that of the fetus in the womb of the mother.
Even though the fetus develops and evolves in every
possible way before it reaches its full development,
from the beginning it belongs to a distinct species.

The same holds true in the vegetable kingdom, where we observe that the original and distinctive character of the species does not change, while its form, colour, and mass do change, transform, and evolve.

To summarize: Just as man progresses, evolves, and is transformed from one form and appearance to another in the womb of the mother, while remaining from the beginning a human embryo, so too has man remained a distinct essence—that is, the human species—from the beginning of his formation in the matrix of the world, and has passed gradually from form to form. It follows that this change of appearance, this evolution of organs, and this growth and development do not preclude the originality of the species. Now, even accepting the reality of evolution and progress, nevertheless, from the moment of his appearance man has possessed perfect composition, and has had the capacity and potential to acquire both material and spiritual perfections and to become the embodiment of the verse, "Let Us make man in Our image, after Our likeness."[26]

6-8 years after the talks recorded in *Some Answered Questions*, 'Abdu'l-Baha made a trip to America. One of the many talks he gave was to a gathering of agnostics at the Open Forum in San Francisco on 10 October 1912. The following is the English translation recorded in *Promulgation of Universal Peace*. See Appendix C for Keven Brown's provisional translation from the original Persian. 'Abdu'l-Baha is reported to have said,

... Between man and the ape, however, there is one link missing, and to the present time scientists have not been able to discover it. Therefore, the greatest proof of this western theory of human evolution is anatomical, reasoning that there are certain vestiges of organs found in man which are peculiar to the ape and lower animals, and setting forth the conclusion that man at some time in his upward progression has possessed these organs which are no longer functioning but appear now as mere rudiments and vestiges.

… The philosophers of the Orient in reply to those of the western world say: Let us suppose that the human anatomy was primordially different from its present form, that it was gradually transformed from one stage to another until it attained its present likeness, that at one time it was similar to a fish, later an invertebrate and finally human. This anatomical evolution or progression does not alter or affect the statement that the development of man was always human in type and biological in progression. For the human embryo when examined microscopically is at first a mere germ or worm. Gradually as it develops it shows certain divisions; rudiments of hands and feet appear—that is to say, an upper and a lower part are distinguishable. Afterward it undergoes certain distinct changes until it reaches its actual human form and is born into this world. But at all times, even when the embryo resembled a worm, it was human in potentiality and character, not animal. The forms assumed by the human embryo in its successive changes do not prove that it is animal in its essential character. Throughout this progression there has been a transference of type, a conservation of species or kind. Realizing this we may acknowledge the fact that at one time man was an inmate of the sea, at another period an invertebrate, then a vertebrate and finally a human being standing erect. Though we admit these changes, we cannot say man is an animal. In each one of these stages are signs and evidences of his human existence and destination. Proof of this lies in the fact that in the embryo man still resembles a worm. This embryo still progresses from one state to another, assuming different forms until that which was potential in it—namely, the human image—appears...

> The lost link of Darwinian theory is itself a proof that man is not an animal. How is it possible to have all the links present and that important link absent? Its absence is an indication that man has never been an animal. It will never be found.[27]

The writings of Shoghi Effendi

Shoghi Effendi was 'Abdu'l-Baha's grandson and the only person authorized after the latter's death to make authoritative interpretations of scripture. He received a letter in 1946 questioning 'Abdu'l-Baha's statements on human origin. The letter was not from a Baha'i, but a friend of the Faith who took issue with some statements from 'Abdu'l-Baha and contrasted the ideas with those of Darwin. The friend asked how certain statements could be proved, writing that the theory of evolution is "recognized by all authorities", and that it does not depend on the discovery of a missing link. He disagreed with the idea of classifying humans and animals separately, saying that they differ only in degree, not in principle. He also disagreed with statements that man breaks the laws of nature or that nature is devoid of memory (see Appendix G). Shoghi Effendi responded via a secretary* on 7 June 1946:

---

* Though not written in his own hand, such letters were reviewed by Shoghi Effendi and are authoritative.

We cannot prove man was always man for this is a fundamental doctrine, but it is based on the assertion that nothing can exceed its own potentialities, that everything, a stone, a tree, an animal and a human being existed in plan, potentially, from the very "beginning" of creation. We don't believe man has always had the form of man, but rather that from the outset he was going to evolve into the human form and species and not be a haphazard branch of the ape family.

You see our whole approach to each matter is based on the belief that God sends us divinely inspired Educators; what they tell us is fundamentally true, what science tells us today is true; tomorrow may be entirely changed to better explain a new set of facts...

These various statements must be taken in conjunction with all the Baha'i teachings; we cannot get a correct picture by concentrating on just one phrase.[28]

In 1947, in response to a question from an individual, the following was written on behalf of Shoghi Effendi:

There is nothing unreasonable in ['Abdu'l-Baha's] statement you quote, page 220*, "The Promulgation of Universal Peace": He expounds the idea that man

---

*   The quotation from 'Abdu'l-Bahá appears on p. 225 of the 1982 edition and p. 315 of the 2012 edition. It reads: "In the world of existence man has traversed successive degrees until he has attained the human kingdom. In each degree of his progression he has developed capacity for advancement to the next station and condition. While in the kingdom of the mineral he was attaining the capacity for promotion into the degree of the vegetable. In the kingdom of the vegetable he underwent preparation for the world of the animal, and from thence he has come onward to the human degree, or kingdom. Through this journey of progression he has ever and always been potentially man."

was always potentially man, which is just another way of saying the Cause contains the power to produce the effect; in this planned and integrated universe, he was part of the plan from the beginning, so to speak.29

Again, on behalf of Shoghi Effendi, in 1950:

The Baha'i Faith teaches man was always potentially man, even when passing through lower stages of evolution. Because he has more power, and subtler powers than the animal, when he turns towards evil he becomes more vicious than an animal because of these very powers.[30]

Unfortunately, the interpretations from Shoghi Effendi still leave slight ambiguity. For example, the key phrase, "from the outset he was going to evolve into the human form and species and not be a haphazard branch of the ape family" can be read two ways: was man a branch of the ape family but not haphazard, or was he not a branch of the ape family? The next comment acknowledges a conflict with contemporary science and reminds that "what science tells us today is true; tomorrow may be entirely changed", but the reader might also emphasize the prior sentence indicating that "a human being existed in plan, potentially" in a timeless sense.

Whereas the excerpt from 1946 is addressing the controversy of common ancestry, the one from 1947 is addressing a talk by 'Abdu'l-Baha that, in turn, appears to not be addressing human evolution. The mention in 1950 seems to support parallel development, although even that has been disputed.

Apparent meaning

There is an obvious and apparent tension with these statements by 'Abdu'l-Bahá and the current understanding of evolution. This tension is acknowledged by Shoghi Effendi. The next chapter demonstrates that varying interpretations are possible: some have declared the tension a misunderstanding, and others have chosen to ignore it. For practical reasons, I chose to use the phrase "apparent meaning" throughout the book to describe what most people would readily acknowledge from reading the text. This phrasing is not meant to declare that it is the correct interpretation, and it is consistent with how most authors have addressed the issue.

CHAPTER 2

AN EXCHANGE OF VIEWS

Early Baha'i authors

Early Baha'is understood the emergence and evolution of humans as separate from animals. Mason Remey wrote an article published in 1922 in *The Star of the West*, an early American Baha'i periodical. Under a section titled "The Evolution of Man" he mentioned,

> Man was created man, a species apart and
> above the vegetable and animal conditions.[31]

Baha'u'llah and the New Era – an introductory book written by a prominent Scottish Baha'i author under the advice of 'Abdu'l-Baha, mentions the topic of evolution in clear language in 1923:

> The human embryo may at one time resemble a fish
> with gill-slits and tail, but it is not a fish. It is a human
> embryo. So the human species may at various stages of
> its long development have resembled to the outward

eye various species of lower animals, but it was still the human species, possessing the mysterious latent power of developing into man as we know him today...[32*]

George Latimer, an early Baha'i from Oregon, wrote a series of articles in *The Star of the West* on "the rapprochement between science and religion". In 1925, in an article titled "Evolution", he wrote:

The progress of the physical body of man resembles the growth and development of the embryo as it passes by degrees from form to form, until it reaches maturity... It is quite possible, without detracting one iota from man's spiritual greatness over other forms of life, to suppose that at one time man walked on his hands and feet, or had a tail; in fact resembled in outer form the animal... The cherry ripens before the apple, but the priority of the cherry does not indicate that the apple was produced from the earlier fruit of the cherry tree. Likewise, even though man is classed as a primate, having hair, shoulder and pelvic girdles of bone, and vertebraes, characteristic to all other mammals, together with a similarity in the functionings of organs, nevertheless, there is no indication of alteration in the original species. From the beginning of man's existence, he has remained a distinct species. Similarity is not a proof of man's development from one primal stock, or brute ancestry, rather is it open testimony to the universality of the laws of creation.[33]

---

* A 2006 printing of this excerpt added a distancing footnote, "The word 'species'... should not be read with its current specialized biological meaning."

John Ferraby, appointed by Shoghi Effendi to the rank of Hand of the Cause, wrote in his "comprehensive outline of the Baha'i Faith" in 1957,

> [Man] has also the special powers of the human spirit that are denied to the lower kingdoms. Whatever the shape of his body, however far from full physical development he was at any stage of his evolution, he always possessed the human spirit. As compared with the animals of the age, man must always have shown signs of his inner glory... any being able to recognise the signs would have been able to see then as now that man differed from the animal.
>
> This principle has often been stated by 'Abdu'l-Baha, Who asserted that scientists would one day recognise its truth.
>
> ...Maybe the scientists of the future will find that evolutionary development is governed by what the creature is destined to evolve into, as well as by ancestry and environment.[34]

Another Hand of the Cause, Hasan Balyuzi, wrote in a 1971 summary of 'Abdu'l-Baha's talk in San Francisco:

> He rejected the theory that man had his origin in the animal kingdom... Man, 'Abdu'l-Baha asserted, was not descended from the animal kingdom; man had always been potentially man, no matter what forms he had assumed in his previous stages of existence.[35]

Modern Baha'i authors

John Hatcher briefly summarized the commonly-held understanding of evolution in *The Purpose of Physical Reality* (1987).

> According to the Baha'i writings the human being has not evolved from other, lower forms of life. The human physical form has evolved, just as an embryo may at first appear to be a tadpole or may later assume various other forms of life. But only the human embryo will, upon reaching fruition, become a human being. In the same way, the human species may at one time in its evolution have appeared to be similar in form to other species, but the human species has always been a distinct creation.[36]

This view carries an obvious tension with a scientific consensus that is confident of common ancestry. The Baha'i principle of the harmony of science and religion requires some reconciliation of the tension. The first modern attempt at reconciliation came from Anjam Khursheed with *Science and Religion: Towards the Restoration of an Ancient Harmony* (1987). He dedicates a third of the book to the topic of evolution, and describes the Baha'i view as antithetical to a pervasive materialistic view of nature that rejects the human soul. On the topic of whether humans and apes share a recent common ancestor, Khursheed is oddly quiet.

Since the 1990s almost every article and book by individual Baha'is on the subject has been obedient to the perspective that religion must be in accord with current scientific knowledge. From 1990 to 2009 there were at least 19 books and articles from 16 authors trying to address the Baha'i approach to evolution based on 'Abdu'l-Baha's comments (see Appendix A for full list). The majority of them did not question the common ancestry of humans and apes.

Some attempted to reconcile the conflict without invalidating 'Abdu'l-Baha as a religious authority by investigating the context and language in which the views were expressed, to explain that his real intention was in accord with modern scientific understanding, and the apparent meaning is actually an unfortunate semantic mistake.

Some tried to bridge the apparent discontinuities, some addressed the scientific challenges of trying to show parallel evolution, some simply declared that no conflict existed, and some emphasized the deleterious force of social Darwinism, or the reductionist, godless direction that society was moving towards, and viewed 'Abdu'l-Baha's statements as a rejection of those notions.

Baha'u'llah's Teachings on Spiritual Reality, compiled by Paul Lample (1996), contains a footnote from the editor that is typical of how these authors approached the issue:

> The Baha'i teachings uphold the scientific concept of evolution, but reject materialistic assertions that the mechanism of evolution is solely based upon chance. In the Baha'i view the appearance of life – and particularly human life – is the outcome of a predominantly purposeful rather than accidental process. The idea that "man was always man" should not be understood to require a kind of parallel evolution; when the appropriate conditions emerged on the planet, then human life appeared.[37]

Prominent among authors of this period is an article by Brown in *Evolution and Baha'i Belief* (2001). Brown's portion of the book

is a significant contribution to the discourse. He looks at philosophies surrounding evolution and speciation, starting with Plato but mostly dedicated to the early 20th century to provide the social context in which 'Abdu'l-Baha was speaking. Brown is even able to recount the ideas that were prevalent in Palestine, where 'Abdu'l-Baha lived, including articles on evolution published in a magazine that he likely read. Regarding the tension with biological science, Brown suggests that the terms used by 'Abdu'l-Baha "should be properly understood through a careful study of their original context, and then they should be interpreted and applied in terms that make sense today."[38] Brown provides a few ideas that could interpret 'Abdu'l-Baha's talks in a way that is consistent with common ancestry, but also believes that 'Abdu'l-Baha rejected the transmutation of species, considered "Darwinian or inter-species evolution"[39] to be an error, and intended "his words on this subject to be taken at face value… with unambiguous and non-symbolic language."[40] He concludes,

> 'Abdu'l-Baha's response to Darwinism was more philosophical in nature than scientific and his main objective was to establish by *rational arguments* the existence a divinely ordained purpose for life, the special place of humanity in creation, the need of final causes (i.e. teleology), and the existence of timeless natural laws in the universe.

The human essence

By far the most common resolution to the tension, as demonstrated by authors from 1990 to 2009, places emphasis on the "essence" or "spirit" of mankind rather than the species. If the term "human" is applied to the point in evolution where *Homo sapiens* emerged (maybe ~300 k years ago), then there is no conflict.

In this view, 'Abdu'l-Baha was speaking not of evolutionary relationships but rather of potentiality and human qualities. He said, "You cannot apply the name 'man' to any being void of this faculty of meditation; without it he would be a mere animal..."[41]

Reading 'Abdu'l-Baha's statements in this way, one could conclude that the evolutionary process was destined to generate a creature capable of pondering its own emergence. Once that happened, the creature had a material lineage from the animal, but attracted the eternal human spirit that was always potential, even inevitable. Human distinction is defined by intellectual powers that are not material, so in that sense they don't "come from" an animal and humans were never animals. For example, in *Paris Talks* 'Abdu'l-Baha is reported to have said,

Although man is part of the animal creation, he possesses a power of thought superior to all other created beings.[42]

Man – the true man – is soul, not body; though physically man belongs to the animal kingdom, yet his soul lifts him above the rest of creation.[43]

And in *The Promulgation of Universal Peace*,

All the powers and attributes of man are human and

hereditary in origin – outcomes of nature's processes
– except the intellect, which is supernatural.[44]

This emphasis on the eternal human spirit is then applied across all references to "man" when discussing evolution. For example, "man has never been an animal" is read to indicate that the "reality" or "essence" of man is not material, nor derived from an animal lineage.

The authors proposing this new viewpoint often review the Platonic philosophy of universals and forms. With this as a background, the authors indicate that the eternal human "essence" made its appearance when one particular branch of the evolutionary tree began to manifest a soul, a lineage that moved from one essence to another, though the essence did not change. The idea of a human essence is well supported in the writings of 'Abdu'l-Baha (see Chapter 7).

This approach was clearly described in Craig Loehle's *On the Shoulders of Giants* (1994) which dedicates a chapter to the Baha'i perspective on evolution. In clear language, Loehle describes modern humans as "different from the animals in kind through possession of a soul but linked to the animals by lineage and physical attributes".[45] So, according to Loehle, just as the human soul exists latent in an embryo, so the human soul  was latent in the mill of evolution until it became manifested in a form that could express human qualities. While this is entirely reconciled with common ancestry, he also adds that subtle, supernatural interventions guided the evolutionary process,

I postulate (the Baha'i writings do not specify this) that

divine Will may have operated at times to help guide the process towards humanity; it was God's intention from the beginning that humanity should arise.[46]

The emphasis on the eternal human essence in 'Abdu'l-Baha's writings provides an avenue for interpreting his comments with social and spiritual implications. The adoption of this view by a majority of Baha'i writers indicates that belief in the Revelation of Baha'u'llah does not require a disbelief in universal common ancestry or any other scientific doctrine.

Infallibility

In 2009 Salman Oskooi presented a thesis to the faculty of San Diego State University on the subject of the harmony of science and religion as applied to biological evolution in Baha'i belief. He noted that the various Baha'i authors have been "unsuccessful in harmonizing this particular set of religious beliefs with the latest science on evolution" and he proposed a different approach of simply acknowledging that 'Abdu'l-Baha was fallible on scientific matters.[47]

Infallibility is central to the discussion. Baha'u'llah is the most recent appearance in a series of prophetic Messengers of God that includes Muhammad, Jesus, and others. His life and teachings are a source of truth, bringing principles that transcend and encompass all areas of human knowledge. Baha'u'llah conferred authority on his son 'Abdu'l-Baha as the person to whom all Baha'is should turn after his own passing for guidance and interpretations. Could "conferred" infallibility not extend to scientific matters? If not, then there would be no conflict, just a recognition that 'Abdu'l-Baha erred.

Oskooi addressed the magnitude of the dilemma of common

ancestry, asserting that the meaning of 'Abdu'l-Baha's statements cannot be explained away by reinterpreting context and language. His personal study of the writings of 'Abdu'l-Baha concluded that they have an "apparent discord with science", "appear uninterpretable in any sense but their apparent meaning", and the apparent meaning is that "humans have been distinct from other beings since the time of some primitive stage of our evolution", without "physical ancestry with animals".

The thesis went as far as to investigate the strength of the evidence for common ancestry, concluding that,

Few, if any, scientific arguments can be made to support the implications of 'Abdu'l-Baha's statements on humankind's unique material origins and intraspecies evolution... Because humans are not merely similar physiologically, but incredibly similar *genetically* to animals that are said to be evolutionarily distinct and separate from them, parallel evolution is not a favored theory... [T]here is no room to doubt that humans share common ancestors with other primate species.[48]

Oskooi's article took an interesting approach to the question of evolution by referencing a letter written on behalf of Shoghi Effendi indicating that his infallibility "is confined to matters which are related strictly to the Cause and interpretation of the teachings; he is not an infallible authority on other subjects, such as economics, science, etc."[49] Oskooi extended this principle to 'Abdu'l-Baha, based on their shared role as interpreters.

He concluded that 'Abdu'l-Baha's statements on human origin conflict with reality, but that this problem does not "contradict the fundamental premises of the faith itself". Ultimately, the Baha'i

Faith centers on Baha'u'llah as an infallible Manifestation of God revealing fundamental truth. The inerrancy associated with such a figure does not necessarily extend to anyone besides Baha'u'llah and the inerrancy conferred on 'Abdu'l-Baha may not extend to scientific matters.

Oskooi's paper provides an option to consider for interested followers of the Baha'i view on evolution. But this approach, ultimately, seems unlikely to resolve the dilemma. Not only does it leave the unpalatable conclusion that 'Abdu'l-Baha said things that are demonstrably untrue, but Oskooi missed something. The Secretariat of the Universal House of Justice responded in 1982 to the question of whether Shoghi Effendi's comment might be extended to 'Abdu'l-Baha:

> There is nothing in the Writings that would lead
> us to the conclusion that what Shoghi Effendi says
> about himself concerning statements on subjects not
> directly related to the Faith also applies to 'Abdu'l-
> Baha. Instead we have assertions which indicate that
> 'Abdu'l-Baha's position in the Faith is one for which
> we find "no parallel" in past Dispensations.[50]

Passages not addressing Darwinian evolution

Oskooi argued that those authors who outright dismiss the apparent tension and declare 'Abdu'l-Baha's comments to be in agreement with modern science are taking quotes out of context and downplaying the most explicit references.

Several such quotes by 'Abdu'l-Baha have been used. They at first appear to support a single origin of all life on earth, but on closer examination they are not discussing evolution of the species. For example, 'Abdu'l-Baha said,

the origin of all material life is one[51]

This has been excerpted by itself without context to suggest common biological origin, but the entire sentence reads, "Verily, the origin of all material life is one and its termination likewise one." Reading the full talk, he is referencing the idea that the human body is made up of the same elements and molecules as make up the bodies of plants and animals.

In another place 'Abdu'l-Baha said that the human body,

has risen by evolution into the kingdom of the animal and from thence attained the kingdom of man.[52]

This appears to give strong support to standard evolutionary theory, but a simple reading of the entire talk reveals that it is not addressing evolutionary relationships, but the composition of the body. The next sentence reads, "After its disintegration and decomposition it will return again to the mineral kingdom".

In yet another example of a quote that has been taken out of context, 'Abdu'l-Baha said,

Consider the world of created beings, how varied and diverse they are in species, yet with one sole origin.[53]

However, the full context is clear that the "sole origin" is God, not a universal common ancestor.

Science can change

Besides Oskooi's thesis, 2009 also saw the publication of 'Some Answered Questions: A Philosophical Perspective', in *Lights of Irfan*, vol. 10, by Ian Kluge. Somewhat unique among the crowded author-space on human origin, this lengthy article advocates in unambiguous language that 'Abdu'l-Baha's comments do, in fact, conflict with current scientific thought. He then suggests that the principle of the harmony of science and religion does not mean that religion must conform to any and all contemporary scientific conclusions:

There is no question that 'Abdu'l-Baha's views on human evolution are in conflict with current scientific thought in regards to the origins and history of humankind. However, this does not necessarily undermine Baha'u'llah's teaching that science and religion should be in harmony unless one adopts the view that religion must uncritically agree with science on all its pronouncements at all times. Logically this is untenable for the simple reason that science itself changes its views — sometime profoundly — and no text, revealed or not, can adopt all the successive scientific beliefs on a given subject without falling into self-contradiction and, thereby, ceasing to be useful as a guide.

Kluge voiced my own conclusion. The discourse trying to resolve the tension over human origin failed to mention a remarkably simple idea that was clearly advocated by Shoghi Effendi: science can change. It is acceptable to live with some ambiguity over what the text meant and what contemporary science has concluded. To insist on a stretched interpretation of 'Abdu'l-Baha as the only valid viewpoint, or to insist on an unobservable ancestor as unassailable truth, seems wrong.

However, science in the West is battling the creationist movement, which taints any notion of deviating the slightest from standard evolutionary theory. I understand the attractiveness of reconciling this issue while adhering strictly to modern science, but my personal reflection on the talks of 'Abdu'l-Baha lead me to believe that he was sharing a principle that has not yet been fully realized by science, a principle that has both biological and social implications. I also understand the slippery slope of contending with science in every instance where one's personal ideas are in conflict with established knowledge – an attitude that should be strictly avoided.

CHAPTER 3

TRANSLATING, INTERPRETING, AND INVESTIGATING

Revised translation

The table talks between Laura Clifford Barney and 'Abdu'l-Baha that make up *Some Answered Questions* were originally recorded in Persian, then reviewed, corrected, and approved by 'Abdu'l-Baha. The English translation by Barney in 1908 was clearly not up to the standard of a modern translating committee, and it wasn't until March 2014 that the fruit of a committee at the Baha'i World Centre was brought to print.

The revised translation raised an entirely new question about the controversy over evolution: could a poor translation be the source of a semantic error that is resolved in the new translation? The short answer is, no. For the long answer, review Chapters 46–51 in both translations (Appendix B), or see the table below for a few examples.

Ch.para.	1908 translation	2014 translation
47.11	Man was always a distinct species, a man, not an animal.	But man has always been a distinct species; he has been man, not an animal.
49.8	To recapitulate: as man in the womb of the mother passes from form to form, from shape to shape, changes and develops, and is still the human species from the beginning of the embryonic period – in the same way man, from the beginning of his existence in the matrix of the world, is also a distinct species – that is, man – and has gradually evolved from one form to another.	To summarize: Just as man progresses, evolves, and is transformed from one form and appearance to another in the womb of the mother, while remaining from the beginning a human embryo, so too has man remained a distinct essence – that is, the human species – from the beginning of his formation in the matrix of the world, and has passed gradually from form to form.
49.8	Man from the beginning was in this perfect form and composition, and possessed capacity and aptitude for acquiring material and spiritual perfections…	…from the moment of his appearance man has possessed perfect composition, and has had the capacity and potential to acquire both material and spiritual perfections…

Ch.para.	1908 translation	2014 translation
51.3	So also the formation of man in the matrix of the world was in the beginning like the embryo; then gradually he made progress in perfectness, and grew and developed until he reached the state of maturity, when the mind and spirit became visible in the greatest power. In the beginning of his formation the mind and spirit also existed, but they were hidden; later they were manifested.	Likewise, at the beginning of his formation in the matrix of the world, man was like an embryo. He then gradually progressed by degrees, and grew and developed until he reached the stage of maturity, when the mind and the spirit manifested themselves in the utmost perfection. From the beginning of his formation, the mind and spirit existed, but they were hidden and appeared only later.

While a few isolated quotations may be translated slightly in favor of reconciliation with the 'human essence' interpretation, the overall picture of tension remains the same.

The tension is directly addressed in the Foreword to the 2014 edition by an unnamed author, clearly summarizing what has been the standard approach to the issue by modern Baha'i authors:

> Religious belief should not contradict science and reason. A certain reading of some of the passages found in Chapters 46–51 may lead some believers to personal conclusions that contradict modern science. Yet the Universal House of Justice has explained that Baha'is strive to reconcile their understanding of the statements of 'Abdu'l-Baha with established scientific perspectives, and therefore it is not necessary to

conclude that these passages describe conceptions rejected by science, for example, a kind of "parallel" evolution that proposes a separate line of biological evolution for the human species parallel to the animal kingdom since the beginning of life on earth.

A careful review of 'Abdu'l-Baha's statements in this volume and in other sources suggests that His concern is not with the mechanisms of evolution but with the philosophical, social, and spiritual implications of the new theory. His use of the term "species", for example, evokes the concept of eternal or permanent archetypes, which is not how the term is defined in contemporary biology. He takes into account a reality beyond the material realm. While 'Abdu'l-Baha acknowledges elsewhere the physical attributes that human beings share in common with the animal and that are derived from the animal kingdom, in these talks he emphasizes another capacity, a capacity for rational consciousness, that distinguishes man from the animal and that is not found in the animal kingdom or in nature itself. This unique capacity, an expression of the human spirit, is not a product of the evolutionary process, but exists potentially in creation... His essential argument, then, is not directed towards scientific findings but towards the materialist assertions that are built upon them. For Baha'is, the science of evolution is accepted, but the conclusion that humanity is merely an accidental branch of the animal kingdom – with all its attendant social implications – is not.[54]

This Foreword hints at a somewhat new perspective. 'Abdu'l-Baha's own emphasis on science and reason may trump any perceived conflict. He said, "Every religion which is not in accordance with established science is superstition."[55] It is possible that Baha'is may simply declare standard evolutionary theory to be correct because it is established science. Maybe the Foreword provides the necessary interpretation for the time.

The book was published by the Baha'i World Centre and the Foreword is unauthored. In response to a question about its authority, the Secretariat responded that the paragraph in the Foreword regarding evolution was "approved for inclusion" by the House of Justice.[56]

The authority of the Foreword is nuanced. There is a clear threshold for material being the authoritative communication of the Universal House of Justice. Below that threshold there is a category of material that the Universal House of Justice commissioned, approved of, or commended for study, and these are very useful for gaining an enhanced understanding of the Baha'i Teachings and putting them into action. Publications such as *One Common Faith* or *Century of Light*, memoranda from the Research Department, or the social action paper prepared by the Office of Social and Economic Development at the Baha'i World Centre are in this category. They provide insights, but they represent the views of those departments and authors and are not binding for Baha'is in the way that messages from the Universal House of Justice are.

Interpretation of the Baha'i writings

There is a reason why the Foreword of *Some Answered Questions* cannot be understood as offering an authoritative interpretation. Regarding interpretation of scripture, there is a subtle and important nuance in Baha'i doctrine. Authoritative interpretation of Baha'i writings after 'Abdu'l-Baha was the exclusive right of Shoghi Effendi, while the Universal House of Justice has the exclusive right to legislate on matters unaddressed in scripture. "Neither can, nor will ever, infringe upon the sacred and prescribed domain of the other", according to Shoghi Effendi.[57]

The Universal House of Justice likewise confirmed that it "cannot and will not infringe upon that domain"[58] and that it "will not engage in interpreting the Holy Writings."[59] However, it is also charged with deliberating on "all problems which have caused difference, questions that are obscure and matters that are not expressly recorded in the Book",[60] and has a role in providing vision and direction, and maintaining unity of thought and action.

So how can the Universal House of Justice maintain unity among inevitably varying interpretations of scripture while itself not providing authoritative interpretations? Among other things, this was addressed in two major messages from the Universal House of Justice, in 1965 and 1966.[61]

The first letter confirmed that "authoritative" or "inspired" interpretations are prohibited after Shoghi Effendi[62] and that the Universal House of Justice will only operate within a sphere of jurisdiction that Shoghi Effendi described as "clearly defined".[63] It wrote that unity of doctrine is maintained by scripture and the "voluminous" interpretations of 'Abdu'l-Baha and Shoghi Effendi, while unity of administration is "assured" by the authority conferred upon the Universal House of Justice and its ability to adapt to the needs of a changing society.

The second letter further elucidated how to navigate the situation,

> If some of the statements of the Universal House
> of Justice are not detailed the friends should
> realize that the cause of this is not secretiveness,
> but rather the determination of this body to
> refrain from interpreting the teachings...

A clear distinction is made in our Faith between authoritative interpretation and the interpretation or understanding that each individual arrives at for himself from his study of its teachings. While the former is confined to the Guardian, the latter, according to the guidance given to us by the Guardian himself, should by no means be suppressed. In fact such individual interpretation is considered the fruit of man's rational power and conducive to a better understanding of the teachings, provided that no disputes or arguments arise among the friends and the individual himself understands and makes it clear that his views are merely his own... although individual insights can be enlightening and helpful, they can also be misleading. The friends must therefore learn to listen to the views of others without being overawed or allowing their faith to be shaken, and to express their own views without pressing them on their fellow Baha'is.[64]

The Foreword to *Some Answered Questions* must be read in light of this guidance: it is not written by or on behalf of the Universal House of Justice; it does not represent an authoritative interpretation; any such interpretation is the view of the author; and any absence of commentary on human evolution by the Universal House of Justice was intentional. Its inclusion was meant to add to the discourse and be seriously considered. It may be considered a preferred understanding that aligns with contemporary scientific findings in a messy discourse that has lacked any form of statement from the Baha'i World Centre, but it does not limit discussion of differing opinions on the matter. Indeed, the clarifying letter about the Foreword, written by the Secretariat, wrote, "different individuals, using their rational powers to reach personal interpretations of scientific findings and the meaning of Sacred Texts, may come to different conclusions on different questions. This is the inevitable outcome of the independent investigation of truth."[65]

The writings of the Universal House of Justice

Though not directly addressing the topic of evolution, this excerpt from a letter of the Universal House of Justice to an individual in 1968 addresses the balancing between shifting theories and the fixed nature of divine revelation:

> While it may often be the part of wisdom to approach individuals or an audience from a standpoint of current knowledge, it should never be overlooked that the Revelation of the Manifestation of God is the standard for all knowledge, and scientific statements and theories, no matter how close they may come to the eternal principles proclaimed by God's Messenger, are in their very nature ephemeral and limited. Likewise, attempting to make the Baha'i Faith relevant to modern society is to incur the grave risk of compromising the fundamental verities of our Faith in an effort to make it conform to current theories and practices.[66]

Just as revelation is fixed and our understanding of it can be in error, scientific facts are distinct from the conclusions of scientists. Another letter written on behalf of the Universal House of Justice to an individual in 1975 wrote:

> Just as there is a fundamental difference between divine Revelation itself and the understanding

that believers have of it, so also there is a basic distinction between scientific fact and reasoning on the one hand and the conclusions or theories of scientists on the other. There is, and can be, no conflict between true religion and true science...[67]

The Secretariat of the Universal House of Justice has responded to several individual queries on the Baha'i perspective on evolution. One such response on 4 September 2005 stated:

[T]here is no statement prepared by the Universal House of Justice that presents in detail the Baha'i perspective on the debate between creationism and Darwin's theory of evolution.[68]

To another individual seeking clarity the Secretariat wrote on 5 July 2010:

The Baha'i view of evolution is more complex and nuanced than that put forward today by those who present evolution and creation in dichotomous terms. Evolution may be understood as the means set in motion by God through which life changes and unfolds. A Baha'i can strive to reconcile contemporary scientific views with the published statements of 'Abdu'l-Baha, which need not be understood to imply a kind of parallel evolution. Rather, 'Abdu'l-Baha has explained that human life came into existence when the appropriate conditions were established.[69]

When the revised translation of *Some Answered Questions* was published in 2014, another individual requested clarity on the Foreword and received a lengthy response on 21 February 2016. The letter was one of seven selected messages shared that year on the official Baha'i Reference Library and the only letter addressing human origin to be given wide dissemination. It stands out as an important milestone in the saga of how to understand the Baha'i perspective on evolution.

The 2016 letter is nuanced and addresses how to apply the harmony of science and religion to the case of evolution. I recommend reading the full text in Appendix H.

The letter points out that in light of all of the statements of 'Abdu'l-Baha on the subject, it is possible for a Baha'i to disagree "with the materialistic philosophical interpretation of scientific findings—that man is merely an animal and a random expression of nature—without contesting the scientific findings themselves, such as those in genetics which are incompatible with a concept of 'parallel' evolution." It encourages Baha'is to "use all of the relevant texts on the subject as well as the most accurate and reliable picture of reality that science can provide to try to understand what 'Abdu'l-Baha actually is conveying", and to avoid the extremes of dismissing truths in the Revelation or contending with scientific findings in every case where one's personal understanding differs.

The letter recognizes that there have been cases where statements in scripture were confirmed centuries later, and there "may well be statements in the Writings about the material world the veracity of which will be proven by science in future." However, "some scientific statements are accurate and reliable descriptions of reality", and 'Abdu'l-Baha "emphasizes that religious beliefs should be weighed in the light of science and reason".

The letter concludes:

Of course, different individuals, using their rational powers to reach personal interpretations of scientific findings and the meaning of Sacred Texts, may come to different conclusions on different questions. This is the inevitable outcome of the independent investigation of truth. On certain matters, there may for a time be a degree of ambiguity; on others, an exchange of views conducted in a consultative spirit may make the truth evident.

In other words, each individual should apply the principle of the independent investigation of the truth, read the scripture and the best guidance from the study of reality, exchange views in a consultative spirit, and come to personal interpretations that will help guide their lives. In its assessment, applying the principle of the harmony of science and religion, 'Abdu'l-Baha "would not make a statement that contradicts reality". If the concept of parallel evolution is incompatible with the best available science, then it is only reasonable for Baha'is to explore interpretations of scripture that find agreement.

Research into the sciences

In the Qur'an, Muhammad says that the sun runs on a fixed course, and in other places that the sun and moon each float in their own continuous orbits. Some translations also describe the sun and moon as spherical orbs, or the planets as floating through space.[70] The astronomers of the time believed that the earth was the center of the universe, and that anyone challenging their ideas was ignorant. Highly accurate predictions of the movement of planets could be made using the earth-centered model. The description of the fixity, motion, and axis of heavenly bodies was a problem for the Muslim theologians of the time, who had to explain away the references.[71] Even though Pythagoras and Plato had established that the earth moves around the sun, this was forgotten until Galileo and Copernicus established the same conclusion in the 16th century, nine centuries after the Qur'an was recorded. Then it became evident that the Qur'an agreed all along with observable facts, and the Muslim theologians then embraced the apparent meaning of the Quranic references.

So far the "exchange of views conducted in a consultative spirit" has focused on exploring the language and context of 'Abdu'l-Baha and finding an interpretation that is not in conflict with textbook science.

Modern Baha'is may be caught in the same predicament that Muslims were in for 900 years. That is, some may be explaining away the apparent meaning of their text because it conflicts with established science. Unless and until science updates itself with better information, it is reasonable to simply adhere to the conclusions of experts regarding evolution. However, simply believing the conclusions of others is antithetical to the independent investigation of truth, a cornerstone of Baha'i thought. 'Abdu'l-Baha's emphasis was that religion should agree with *reason*, not necessarily what is published in scientific journals. So there is also no fault to be found in investigating the reality of evolution for oneself.

Among all the discourse on the Baha'i teachings on evolution there remains an overlooked and scary corner of the debate room that few Baha'is have approached because creationists also hang out there. Maybe the consensus is… wrong. Little mention has been made of a potential refinement of scientific knowledge that shifts the consensus toward the independent descent of humans. What is there to mention? The case for universal common ancestry is practically impenetrable, and any author proposing a true paragraph shift in science would apparently have nothing to discuss other than speculation that things could change in the future.

Is there room to maintain that the apparent meaning might be validated by new scientific discoveries? Could the field of evolutionary biology be subject to impressive models that make highly accurate predictions, but were nevertheless built on a misguided premise?

The Challenge of Baha'u'llah (1993) by Gary Matthews includes a section on evolution that demonstrates a reasonable approach to the issue. His main point is that "the apparent contradiction is nothing more than a question of semantics: perhaps 'Abdu'l-Baha is merely dating man's beginning as a distinct species from the soul's first appearance, to emphasize that we do not derive our higher spiritual nature from our animal forebears".[72] But Matthews also points out that the record of recent human fossils does not show a common ancestor with two clear paths to modern apes and humans. He says that eighty years of intense research have not found fossil proof of kinship and it "no longer seems farfetched to believe 'Abdu'l-Baha was right… the correspondence so far between 'Abdu'l-Baha's prediction and actual events is nothing less than astounding".[73] He concludes that "further research" of both 'Abdu'l-Baha's intended meaning and on the science of human origin "clearly would be appropriate".[74]

Keven Brown's contribution represents in-depth research into the context and intended meaning of 'Abdu'l-Baha's statements, but research into the sciences is so far missing from Baha'i authors to date. Investigating a biological basis for the apparent meaning of 'Abdu'l-Baha's statements raises some difficult questions about the roles of science and religion.

A new set of facts

Similar to Oskooi, I made my own investigation into the evidence for common ancestry to test whether the apparent meaning of independent ancestry has any basis in reality. Where he found "no room to doubt" common ancestry of humans and apes, I found a new set of facts being generated by prominent researchers. I found an understanding of the available data that can support the independent descent of many species, and significant research that shows mainstream science evolving toward a paradigm shift on the origin of species. My original intention was to simply lay out a hypothesis of how parallel evolution might be proven in the future. After making predictions, I found that most had already been discovered.

These new understandings, which have mostly come about since 2000, can be synthesized into a model that may plausibly bring standard science into an agreement with 'Abdu'l-Bahá's apparent meaning of independent descent. The model is supported by recent discoveries and new interpretations of data that have shaken up evolutionary thought among mainstream scientists. It is also speculative, and easily ignored until certain tenets being debated among biologists become more widely adopted.

I share these ideas understanding that a reasonably intelligent person should at this point in the book reject my conclusions. These results came as a surprise to me, as they should to anyone familiar with the absolute dominance of the modern evolutionary paradigm.

Assumptions

To evaluate multiple origins I had to translate 'Abdu'l-Baha's comments into a modern hypothesis, and in doing so I made three reasonable assumptions that were not derived directly from 'Abdu'l-Baha.

First, I assumed the "beginning" of a line of ancestry begins with visible multicellular life, not microorganisms.

Second, I assumed he was speaking about ancestry in parent-child relationships, not other forms of DNA exchange. 'Abdu'l-Baha spoke on the subject 40 years before the discovery of DNA and focused on the potentiality found in seeds and embryos. The importance of this assumption will be clear in later chapters.

Third, I assumed his comments on the originality of humans apply widely to other species as well. So, for example, all species of foxes may share common ancestry, but their line of descent is independent at some level from the great majority of species. This contrasts with other Baha'is, such as Bahman Nadimi, who wrote in 2004 that he interpreted 'Abdu'l-Baha's comments as separate lineages for plants, animals, and humans.[75]

Summary

The following chapters document the results. These fall into several broad categories: a new description of the base of the tree of life and the significance of horizontal gene transfer (Chapter 4), the role of convergence (Chapter 5), evolutionary experiments (Chapter 6) fossils of the human line (Chapter 8), and the incredible amount of data coming from a wave of genetic studies (Chapter 9).

As investigators tried to describe the base of the tree of life, they found that the tree model of descent is less and less applicable to unicellular life. Bacteria do not use sexual reproduction and can share genetic material horizontally between cells. The respectable tree model that works quite well to describe family history over thousands of generations of vertebrates starts to break down while trying to describe the trunk. The tree of evolution is not the standard maple, it is more like a banyan with many root systems, many trees growing from common roots, and branches growing into themselves like a web.

Banyan tree at the home of Thomas Edison. Source: Wikimedia.

Getting back to the foundational question of evolution – the origin of species – there are several proposals and evidence for the independent and parallel growth of many categories of plants and animals out of a network of gene-sharing unicellular roots. There is a species threshold, a point in evolution before which organisms do not share a common ancestral species. Cellular complexity, sexual reproduction, multicellularity, and centralization of the nervous system appear to be major steps that would create a new independent lineage. These processes have produced several lineages and may be ongoing. The banyan tree of evolution may be more like an aspen, a vast root system that creates hundreds of trees over a large area.

Major evolutionary steps were not singular improbable events. Instead, they were driven by natural processes that are repeatable. Nature selects the features and forms that provide the best fitness. The variation may involve chance, but the outcomes are far from random. Natural selection has repeatedly created the same solution to the same problem. A new cast of evolutionary biologists are finding that these evolutionary convergences are prolific, hinting that evolution does not stumble blindly through a space of limitless possibilities. The new view of determinism creates a sense of predictability in evolutionary outcomes and changes the paradigm of evaluating relatedness based on similar characteristics.

Biologists study organisms, and evolutionary biologists study history. There are two key methods of peering into nature's past: fossil evidence and genomic data.

Working backwards from modern humans, fossils show a lineage of upright-walking humanoids going back at least 7 million years and confounding many older assumptions about their relationship to chimpanzees. Meanwhile, very few fossil fragments exist from chimpanzees or gorillas.

Regardless of any physical or fossil analysis, the findings of genetics appear to irrefutably point to common ancestry, because humans and chimpanzees are similar genetically. This is not a simple subject to address, but DNA analysis, long viewed as the ultimate judge of evolutionary relationships, has changed significantly since 2005 as the price of sequencing a genome dropped to a tiny fraction of what it used to be. The cost of unravelling the human genome from 1990 to 2003 was approximately US$2.7 billion. The same cost was $20 million in 2005, $8,000 in 2011, $1,000 in 2017, and $300 in 2020.

Some earlier assumptions about genomes are changing, such as the extent of convergence, functionality, constraint, the role of viruses, or the cell's ability to manipulate its DNA. For example, only in 2014 researchers found that the majority of human DNA is functional, not the evolutionary refuse that had been assumed existed outside of actual genes. Earlier studies assumed that similar "junk" DNA was proof of common ancestry, but when function and purpose can be attributed, the evidence for ancestry becomes more vague.

Among the patterns found in the noisy mountain of genetic data, a handful of small primitive creatures have been found with genetic signatures of vertebrates. These are difficult to explain within the paradigm of universal common ancestry. However, genomic potential in primitive forms is the predictable outcome of a new paradigm, one that sees a multitude of species emerging from a common evolving genetic pool, each new species with increasingly constrained outcomes as it grows in complexity, and selective pressure convergently driving species into preexistent patterns.

Not creationism

A new hypothesis that includes many tree-like structures of descent, instead of just one, should not be confused with anti-Darwinism. In fact, Darwin's contribution is astonishing for just how accurate he was on such a wide array of topics, far ahead of his time. However, Darwin had his own social context. Biblical creationism was a powerful force in the nineteenth century and much of Darwin's scientific evidence was framed as a refutation of supernatural origin, which is evident from reading his work.

The modern debate is not so different. There are many popular authors (Dawkins, Coyne) that continue writing about evolution as if the only options were natural selection and intelligent design. Unfortunately, ideas challenging the doctrine of universal common ancestry are shared by creationists. Creationists, for their part, have written persuasively about evolution by arguing against a set of misunderstandings about evolution. Within this dichotomy, a superficial understanding of the issue would lump my research together with those of creationists. Unlike creationism, the model of parallel evolution does not invoke any supernatural intervention.

My understanding of 'Abdu'l-Baha's description of human origin means that the basic idea of a tree of life with diverging branches is correct, but that it happened more than once, providing a model that might be called parallel evolution, polygenesis, or independent descent. This conflicts with the principle of universal common ancestry, a pillar of Darwinism, but does not conflict with the major pillar of natural selection, or the related idea of descent with modification. The model would rely on entirely natural processes and need confirmation by further scientific inquiry.

There is no need to invoke a supernatural event or faith, so the model of tree-like descent from a common root system is entirely within the realm of testable science, where creationism (a.k.a. intelligent design) is not.

CHAPTER 4

THE BASE OF THE
TREE OF LIFE

Ubiquitous life

The topic of how species evolve is closely connected to the more fundamental question of how life began.

On earth, a creature evolved that could ponder the meaning of its own consciousness. Is that something natural? Did the first self-replicating molecules come from non-living materials? Did they arrive on meteorites? Was it unlikely? The answers to these questions will help answer a question that Star Trek fans have pondered most of their lives: How prevalent is life in the universe?

Baha'u'llah made the dramatic statement that every fixed star has planets and that every planet has "creatures".[76] The average researcher may say that a theory of ubiquitous life is untestable, but in general most would disagree that every planet has some form of life, leaving some tension and ambiguity as to how to interpret the reference. At face value, interpreting "creatures" as living things on planets throughout the universe, would imply an entirely new way of thinking about biology. The key to unlocking this mystery is to investigate how life gets its start.

Many great thinkers have investigated the transition from non-living things into living things. The investigation is about trying to describe natural laws, because the only other options seem to be supernatural or alien origin, and the alien origin does not address how the meteorite/spacecraft received the complex life.

Few of those great thinkers have asked the perhaps more relevant (and testable) question of, how many times did it happen on earth? If rare, then perhaps it happened only once in the Milky Way, and its seeds spread easily between solar systems. If common through simple, natural, and repeatable pathways, then nearly every planet and solar system may have hosted multiple geneses of life. Understanding how life began would shed light on the prevalence of life throughout the universe. Although it appears untestable, the theory of ubiquitous life can possibly be confirmed without leaving the solar system.

The simple and mindless laws of physics just push particles and atoms around according to equations. Chemical affinities cause atoms to form molecules that eventually begin self-replicating into life. Everything before that transition is chemical, so the origin of life is a chemistry problem! Oddly enough chemists will provide the greatest insights into abiogenesis, the natural origin of living things from non-living material.

Diverse cellular life appeared on earth as soon as the environment was hospitable, and evidence for cellular life dates to the oldest available rock layer. If there is a pathway for chemical affinity to form simple biology, then simple biology formed more than once. This is usually confirmed as a thought experiment by scientists. For example, David Raup and James Valentine questioned the view that the origin of life was "exceedingly improbable" and conclude that "multiple origins of life in the early Precambrian is a reasonable possibility", even if all current life were descended from only one such origin. They estimate that life originated at least 10 times on earth.[77]

The current lack of clarity on such an important topic leaves a fascinating area of study for enthusiastic researchers.

Abiogenesis

Many scientists have formed hypotheses and tested those pathways for the natural origin of living things. Since the 1980s, one generally accepted hypothesis was the early earth as a primordial soup of RNA, which turn DNA into proteins, but RNA are too specialized to have been created from chemical processes, so people continued to study the pre-RNA route.

There are theories and experiments of the spontaneous creation of simple cellular life in high levels of heat, electricity, and pressure. Experiments have produced all RNA nucleobases,[78] the building blocks of nucleic acids such as RNA and DNA, but how those turn into cellular functioning is still a bit of an unsolved mystery. Those building blocks have been shown in other experiments to self-assemble into the helix of DNA "in plain water, with no catalysts, and at room temperature"[79] and those naturally-assembled DNA fragments can bond together to form longer chains,[80] all without the aid of biological mechanisms.

The same conditions that have produced these nucleic acid precursors have also produced the materials needed to make natural amino acids and lipids, the other important components of cellular functioning.[81]

This research is amazingly close to filling in all the gaps of the dawn of life. In 2022, two independent labs created RNA molecules that replicate, diversify, and develop according to natural selection. The first to publish was at the University of Tokyo, where they witnessed the transition from chemistry to biology and its adaptation to the environment based on mutations.[82] Next, researchers at the Foundation for Applied Molecular Evolution, Florida, converted precursors into natural RNA by using volcanic glass as a catalyst.[83] The synthesized molecules were stable for months and lengthy enough to grow by natural selection.

Although it does not relate to the origin of species, cellular life logically arose multiple times on earth, and its generation may be ongoing. New strands of self-replicating RNA would likely have gone extinct due to already-filled niches or their inability to meld due to chiral chemical preference, but some would have joined or adapted to the chaotic system of development already ongoing.

Aliens

As exciting as the earthly biogenesis is, the greatest insights will be drawn from the analysis of life found in other parts of the solar system, especially if it formed naturally and is not related to earthly life. Only by comparing two independent systems can the universal and fundamental elements of organic life be distinguished from the arbitrary and stochastic.

Finding life in the oceans of Jupiter's moons could prove that a second genesis took place and that life is an inevitable outcome of the universe. With current trends of space exploration, such life could be found and analyzed in the next several decades. Even one microscopic example could cause a true paradigm shift in biology.

Adaptation

Even if life had multiple origins on earth, earthly life would not survive on most planets. Confirming widespread life in the universe would require demonstrating that life is far more adaptable than imagined and could survive in extremes of pressure or temperature, and with novel chemical compositions. Although they do not demonstrate proof of widespread life, earthly examples in the extremes provide some insights into the adaptability of life.

Life can thrive without energy from the sun. The only known forms of life were driven by photosynthesis, until 1977 when life around deep-sea vents were discovered that worked almost entirely on chemosynthesis (deriving organic material from minerals). Bacteria in the vents use compounds that are highly toxic to most organisms and live in extreme temperature and pressure. Similarly, cave ecosystems rely on chemosynthetic microorganisms that supply the organic material for underground life.[84]

Life can thrive in sub-freezing temperatures. The search for extraterrestrial life is often limited to worlds that contain liquid water. But there are ice worms (*Mesenchytraeus* spp.) on earth that live in glaciers and turn into goo if raised to 5°C. Even the worms' method of mobility in the ice remains a mystery.[85]

Life can thrive in the depths of a planet, with high temperatures and low oxygen. In 2011, the deepest-living land animal was found in a South African mine at a depth of 1.3km. Two species of worms were found at the "impossible depth" and surprised experts.[86]

Life can survive the radiation and deprivation of space. Earthborn *Streptococcus mitis* bacteria managed to survive on a camera left on the moon for three years until it was returned under sterile conditions and examined. The Apollo 12 commander commented, "I always thought the most significant thing that we ever found on the [whole moon] was that little bacteria who came back and lived and nobody ever said [anything] about it."[87] As surprising as were the bacteria, in 2007 a microscopic creature called a tartigrade became the first documented animal to ride the winds of solar radiation in freezing temperatures with no oxygen, and live to tell the tale to its grandchildren.[88]

The trend of these discoveries has been a continual expansion of what is known to be possible. A limited view of biology comes from limited data. Humans have hardly finished exploring their own little planet. If there were creatures swimming in the ammonia oceans of Jupiter, humans might not find them for thousands of years while assuming that liquid water is required for life. Meanwhile, the Jupiter-whales may spend that time discussing how life couldn't possibly form on those tiny, rocky, inner planets, especially the one filled with toxic water and oxygen, devoid of the high pressures and ammonia oceans necessary for life.

Though the text has not been authenticated and is not currently published, pilgrims' notes record 'Abdu'l-Baha making this dramatic statement about the adaptation of life in other parts of the universe.

The earth has its inhabitants, the water and the air contain many living beings and all the elements have their nature spirits, then how is it possible to conceive that these stupendous stellar bodies are not inhabited? Verily, they are peopled, but let it be known that the dwellers accord with the elements of their respective spheres. These living beings do not have states of consciousness like unto those who live on the surface of this globe: the power of adaptation and environment moulds their bodies and states of consciousness, just as our bodies and minds are suited to our planet. For example, we have birds that live in the air, those that live on the earth and those that live in the sea... Beings who inhabit those distant luminous bodies are attuned to the elements that have gone into their composition of their respective spheres.[89]

Replicating how life began, demonstrating the great adaptability of life on earth, or finding alternative geneses of life in the solar system could provide support for Baha'u'llah's statement about widespread creatures without having to visit every planet in the universe to check for bacteria. In the meantime, and without definitive proof, it is reasonable to believe that widespread and well adapted life can be demonstrated in the future.

Cell evolution

Life may have first formed anywhere from 4.2 to 3.5 billion years ago (bya) in deep sea vents, near the surface of the ocean, on a volcano, on a glacier, in a drying puddle, on pumice, in a high phosphorus lake, or in a meteor blast.

During the next two billion years, life existed as prokaryotes (single cells without nuclei). These were simple, small cells, commonly known as bacteria. The ocean was a genetic soup, with cells replicating, mutating, sharing, and fornicating, because cells at that time lacked sexual reproduction. Viruses played an important role in exchanging DNA, but cells also have a method of transferring genetic information horizontally. Whatever attributes proved beneficial to an organism were easily transferred and propagated in other organisms. In such a world, genetic descent is more akin to a web or thicket than a tree.

The next major milestone came between 2 and 1.5 bya with the appearance of eukaryotes, still single cells but with interacting parts inside a membrane. Included in those parts was the nucleus and mitochondria, each with their own storage of DNA, suggesting that the eukaryotic cell is a symbiosis of two previously independent cells. Mitochondria act like power plants and allow eukaryotes to support 200,000 times more genetic information than bacteria, allowing the nucleus to form and produce complex life. This overturns the idea that the jump to complex eukaryotic cells simply required the right kinds of mutations.[90]

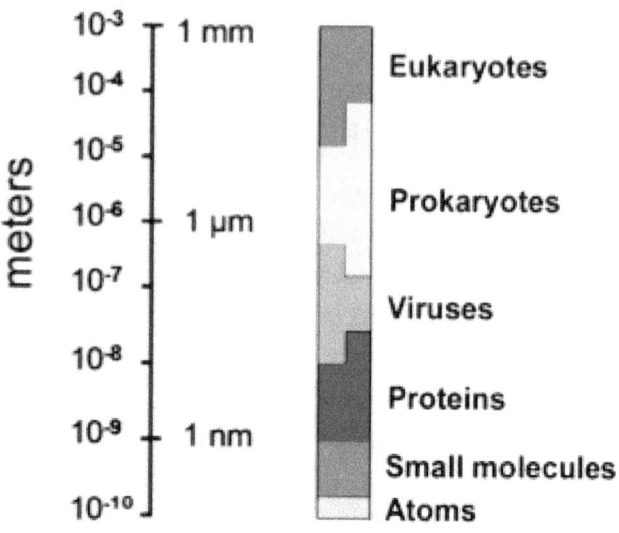

Significant markers in eukaryote evolution correlate to markers in the accumulation of free oxygen in the atmosphere. When oxygen-producing cells saturated the oxygen-capturing matter on earth, excess oxygen began to accumulate in the atmosphere and created mass extinctions of life. Cells that survived were ones that thrived in oxygen. The environment was influenced by organic life. The sensitive balance favorable to life on earth was partly generated by that same life.

Beyond common descent

Perhaps the most accomplished modern researcher into phylogenetic trees, horizontal gene transfer, and cell evolution was microbiologist Carl Woese, recipient of the National Medal of Science, who discovered and elucidated the kingdom of archaea, a third form of life beyond bacteria and eukaryotes. He is perhaps the most important biologist of the 20th century. The evolutionary transition from modular cells using horizontal gene transfer to their organization into genealogical trees has been termed by Carl Woese the "Darwinian Threshold". In a 2002 summary article, 'On the evolution of cells', Woese writes,

> On the far side of that Threshold "species" as we know them cannot exist. Once it is crossed, however, speciation becomes possible. The Darwinian Threshold truly represents the Origin of Species, in that it represents the origin of speciation as we know it.
>
> … The cell… can reach a critical point, where a phase change occurs, where a new, higher level organization of the whole emerges. That, I suggest, is what the Darwinian Threshold represents, a hitherto unrecognized phase change in the organization of the evolving cell.[91]

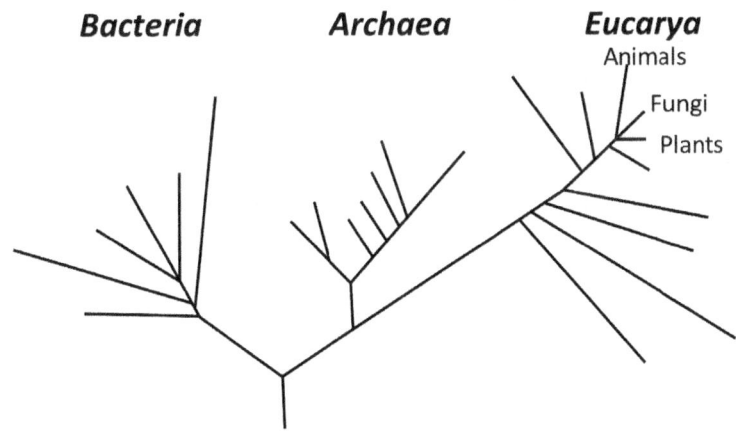

Tree-like structure of cellular evolution. Source: Woese, 2002.

He writes further that the "universal tree has no root", that "classical presumptions do not hold" when approaching the base of the tree, and that what might be considered the first branchings of the universal tree "differ in fundamental ways" and should not be regarded as sister lineages. He summarized the significance of this view:

> Although the archaea and eukarya are represented by a "common lineage" at that stage, this is deceptive: the two are in effect lumped by forcing tree representation on the situation.
>
> ... at the time of the basal bifurcation of the universal tree, the archaeal and eukaryotic cell types were still being communally forged in the chaos of the universal gene-exchange pool, along with other cell types now extinct. Under these evolutionarily fluid conditions, major cellular systems could still be horizontally melded.[92]

In a further 2004 article, he wrote, "what appears in the tree representation to be a common ancestral trunk shared by the archaea and eukaryotes does not actually exist".[93]

This analysis is a major confounding factor to the traditional view of a universal common ancestor. Indeed, Woese made it the object of his research over many decades to describe the attributes of the last universal common ancestor, but concluded that no such organism ever existed.

> From this perspective [of Horizontal Gene Transfer] we will see that there was not one particular primordial form, but rather a process that generated many of them, because only in this way can cellular organization

evolve. The Doctrine of Common Descent (and classical evolutionary thinking in general) rests on the tacit assumption that the dynamic of the evolutionary process remains unchanged as it gives rise to increasingly complex, specific, etc. cellular forms. Yet the forms in essence are the process. Therefore, fundamental changes in their nature can only mean changes in the underlying evolutionary dynamic. The time has come for Biology to go beyond the Doctrine of Common Descent.[94]

Using Darwin's own terminology of "one primordial form", Woese indicates that there were "not one… but rather a process that generated many of them". He goes on to identify three distinct cell types from which all life on earth evolved in genealogical trees that do not share a common ancestral species. The "tree of life" is not one single tree, but three trees (maybe four, with a possible new domain discovered in 2011[95]). As Rachel Moeller, writing in the *Scientific American*, put it: "Instead of one universal evolutionary tree, picture a three-trunk stand sharing a communal root system.[96]

In this sense, the view of *universal* common ancestry has already been overturned and multiple origins from a network of primitive cells is established, with several independent lineages. However, this view does not by itself validate my understanding of 'Abdu'l-Baha's comments, unless multiple origins can be established within cell types, especially within eukaryotes.

Questioning the tree

Scientists form hypotheses, perform tests on observable phenomena to answer questions, but only the questions that scientists are asking. Underlying research for many decades has been an unquestioned assumption that all earthly life evolved from a last universal common ancestor (LUCA) through diverging branches. As biochemists Doolittle and Bapteste put it, "Seldom have investigators asked whether non-tree (reticulated) models might not better explain the data at hand."[97]

Their comment underpins the ideas of many authors investigating the root of the universal tree, the time before visible life appeared. As Woese and others began investigating in more depth, they found that the very idea of a universal tree is inconsistent with reality. As it turns out, the majority had not been asking the right questions until recently. In a 2000 article 'Uprooting the tree of life', Doolittle wrote:

> ... the pattern of evolution is not as linear and treelike as Darwin imagined it... we must now admit that any tree is at best a description of the evolutionary history of only part of an organism's genome. The consensus tree is an overly simplified depiction...

> Though complicated, even this revised picture would actually be misleadingly simple, a sort of shorthand cartoon, because the fusing of branches usually would not represent the joining of whole genomes, only the transfers of single or multiple genes. The full picture would have to display simultaneously the superimposed genealogical patterns of thousands of different families of genes... [T]here would never have been a single cell that could be called the last universal common ancestor.[98]

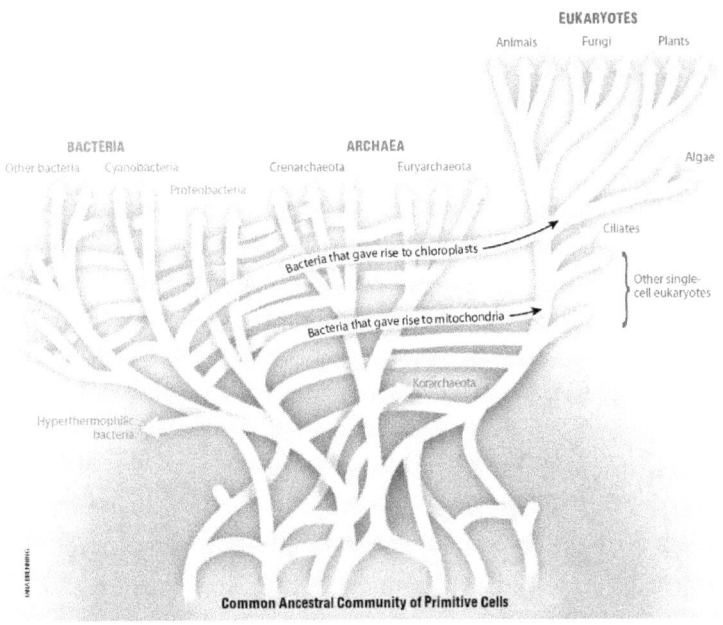

EUKARYOTES

Animals   Fungi   Plants

BACTERIA   ARCHAEA
Other bacteria   Cyanobacteria   Crenarchaeota   Euryarchaeota   Algae

Proteobacteria

Ciliates

Bacteria that gave rise to chloroplasts

Other single-
cell eukaryotes

Bacteria that gave rise to mitochondria

Korarchaeota

Hyperthermophilic
bacteria

Common Ancestral Community of Primitive Cells

*"REVISED 'TREE' OF LIFE retains a treelike structure at the top
of the eukaryotic domain and acknowledges that eukaryotes obtained
mitochondria and chloroplasts from bacteria. But it also includes an
extensive network of untreelike links between branches. Those links have
been inserted somewhat randomly to symbolize the rampant lateral gene
transfer of single or multiple genes that has always occurred between
unicellular organisms. This 'tree' also lacks a single cell at the root;
the three major domains of life probably arose from a population of
primitive cells that differed in their genes." Source: Doolittle, 2000.*

By 2008, Doolittle was joined by other evolutionary biologists
in criticism of the concept of a tree of life – as applied to microbial life. In addition to the newfound prominence of horizontal
gene transfer, they pointed to a much higher rate of hybridization among species than was ever imagined. Among many authors
noting the need for reform of certain axioms in biology, Eugene
McCarthy introduced his book in the following way:

For the last 150 years, we biologists have been defending a fortress built by Charles Darwin. We have spent our energies hurling back the assaults of the creationist infidels and shoring up a slowly crumbling foundation that once seemed based on the hard bedrock of direct observation. But an ocean of data, accumulating since 1859, has been slowly lapping away at the rotting stone beneath Darwin's castle, undermining its moldering walls, making it an ever more dangerous place to reside.[99]

Articles in *The Telegraph*[100] and *The Guardian*[101] in 2009 documented a growing consensus that Darwin's tree of life is "wrong", "misleading", "obsolete", "an oversimplification", and "needs to be discarded", "replaced", or "politely buried" in light of horizontal gene transfer, gene recombination, convergence, hybridization, and a "far more complex scenario than Darwin could have imagined" when trying to describe evolutionary relationships at the base of the tree. Michael Rose of UC Irvine said, "What's less accepted is our whole fundamental view of biology needs to change." Others write, "the belief that prokaryotes are related by such a tree [of life] has now become stronger than the data to support it." Some research since 2005 has used the term "phylogenetic net", "net of life", or even "forest of life" instead of "tree of life".[102]

Nicholas Barton and colleagues wrote a textbook on evolutionary biology in 2007 that read,

…there cannot be a single 'Tree of Life'. That is, a single tree cannot accurately represent the evolution of life. It may be better to represent species evolution as a reticulated network with interconnecting branches.[103]

In 2008 the National Center for Science Education wrote,

> This view of the tree of life with a reticulated network of roots replaces the concept of the last universal common ancestor (LUCA) with the concept of a community of common ancestors who are related to one another via genetic exchanges. Although the reticulated tree of life began as a controversial idea, it is now fully embraced as a plausible evolutionary scenario.[104]

The concept of the last universal common ancestor (LUCA) has been replaced with a community of common ancestors in a chaotic gene exchange pool with probably many roots. Clearly, the old orthodoxy of LUCA is still being repeated in textbooks, but others are on the cutting edge.

Endosymbiosis

Just as with the formation of simple prokaryotes, endosymbiosis – the process that formed eukaryotes – could be repeatable and ongoing. This particular jump in complexity is less studied than the origin of life, but there is evidence for multiple origins of eukaryotes.

A 2005 paper in the journal *Protist* indicated that "the evolution of photosynthetic organelles from cyanobacteria was not a unique event, as is commonly believed, but may be an ongoing process".[105] The same year an article in *Science* suggested that the relationship between an observed protist and green algae represents an endosymbiosis in progress.[106]

A paper in 2008 suggested that the acquisition of plastids "is better described as the result of a process rather than something occurring at a discrete time".[107]

Another study of ancestry in 2008, using the broadest dataset to date, compared genetic elements in 65 species, investigating early evolution among eukaryotes. Among the conclusions, the researchers found "strong support" for separating eukaryotes into two "monophyletic megagroups" with separate eukaryotic origins.[108]

Further research published in 2011 acknowledged that plastids "have had multiple independent origins in different phyla".[109]

In plain English, a deep look at the genetics and physiology of living organisms indicates that nearly all plant life using photosynthesis came from a unique line of ancestry all the way back to the formation of the eukaryote (endosymbiosis). This formation appears to be not a one-of-a-kind event, but a process that has produced many fruits. The 'early' stages of life on earth are being repeated in the modern day.

Plant life has a separate eukaryotic origin, so plants and animals do not share a common ancestor and animal life did not arise from plant life. Any proposed 'common ancestor' would have been several lineages of cells that did not use sexual reproduction. 'Ancestry' is no longer a meaningful way to describe the relationships of those cells or their genomes. If there is a shared genealogical past to plants and animals, it came from the parallel emergence from a community of gene-sharing cells. There would be no 'common ancestor'.

Perhaps more radical, experts suggest that this parallel emergence is not a unique event, but an ongoing process.

Independent descent of plants and animals is certainly in the direction of supporting parallel evolution, but the idea of all animal life descending from a single eukaryotic formation still would conflict with my understanding of 'Abdu'l-Baha. There would need to be many more eukaryotic origins, or maybe there is a different evolutionary transition we should be paying attention to.

Multicellularity

After the formation of the eukaryote, the next major evolutionary step is multicellularity, which is an interesting transition because it represents a phase change, not just more of the same old cellular evolution. If multicellularity arose independently among several lineages, then they would not share ancestry in a fundamental way. Was multicellularity formed many times in independent lineages?

Indeed, according to a 2010 article in *Nature*, it has already been well established that multicellularity was evolved independently in "most major lineages of eukaryotic organisms", not only plants and animals, "but also green algae, brown algae, red algae, ciliates, slime molds, and fungi".[110] The same study notes that multicellularity "can evolve through relatively minor modifications of the unicellular blueprint", hinting that it should not be such a rare process as has been assumed. A separate 2010 study showed that, "The transition from colonies of individual cells to multicellular organisms can be achieved relatively rapidly, within one million generations" (two to five thousand years), using a new mathematical model.[111]

Though multicellularity is known to have developed many times, few people have attempted to recreate it. In a 2011 experiment, two scientists from the University of Minnesota found that achieving multicellularity was actually very easy. Using yeast cells and a centrifuge, in 60 days several strains of yeast cells formed multicellular spheres.[112]

The transition from unicellular to multicellular life turns out to be no big deal. It is a natural and repeatable process. It is established knowledge that it happened numerous times in the history of life. An earlier 2007 analysis by researchers from the University of California and University of Washington noted that evolving multicellularity is an observable and inducible response to environmental stimuli. "Thus, the transition to multicellularity is relatively easy – a minor major transition."[113]

Independent multicellular formations of animals, algae, and fungi is certainly in the direction of supporting parallel evolution, but the idea of all animal life descending from a generalized common ancestor still would conflict with my understanding of 'Abdu'l-Baha. There would need to be many more animal origins.

The species threshold

In the early evolution of a cell, genetic material does not follow a branching pattern of descent as it is copied into numerous organisms. But at some point along the way the cell grows in complexity until its progeny gradually form into a clear trunk with diverging branches.

Carl Woese proposed that this 'species threshold' – a crossing into vertical evolution – is associated with cellular complexity and the development of sexual reproduction. This is on one end of the spectrum of ideas about where the threshold sits for the formation of a species with branching lines of descent. On the other end, others have focused on the centralization of the nervous system. Yet others place emphasis on multicellularity.

I propose that all of these play an important role, but multicellularity is the attribute that should be most associated with the origin of a species. As a multicellular transition stabilizes into a distinct population, it first makes its appearance into visible life and has a distinct starting point. If someone wants to say that two lines of descent share a 'common ancestor' before this point, that is understandable, but it may be using the tree model where it doesn't belong, where horizontal gene transfer and endosymbiosis are still exerting a great influence on genomes.

The remainder of this chapter documents various authors,

independently proposing variations of the theme of a species threshold with multiple origins. This trend should peak the interest of anyone investigating 'Abdu'l-Baha's comments on human origin.

Evolution of development

Stuart Newman, Ph.D., Professor at New York Medical College, has developed an idea of early animal development that adds to our understanding of the origin of species. Published in the 12 October 2012 issue of *Science*, he describes early animal evolution as going through four stages: first a single cell representing a proto-egg form, second a multicellular cluster, third an intermediate form that is a combination of developmental motifs, and fourth a stable lineage representing a specific organism. Newman proposes that, following this model, distinct animal phyla arose in parallel.[114]

Newman's research focuses on the developmental motifs, which he says are based on physics and are similar to nonliving, chemically active materials. The unicellular progenitor of the developing multicellular cluster contains a mix of genes in the developmental genetic 'toolkit'. The options available in the toolkit cause certain patterns of adhesion, folding, layering, mixing, repelling, and elongating that are repeatable and predictable. Depending on the options available to the progenitor, most clusters become non-viable, but some succeed in a few key generic forms.

This process produced several organisms that eventually coalesced into stable forms that further evolved via natural selection. According to Newman,

> the nongeneric mechanisms are evolved embellishments of the generic ones, with selection stabilizing and reinforcing inherent forms rather than inventing new ones.[115]

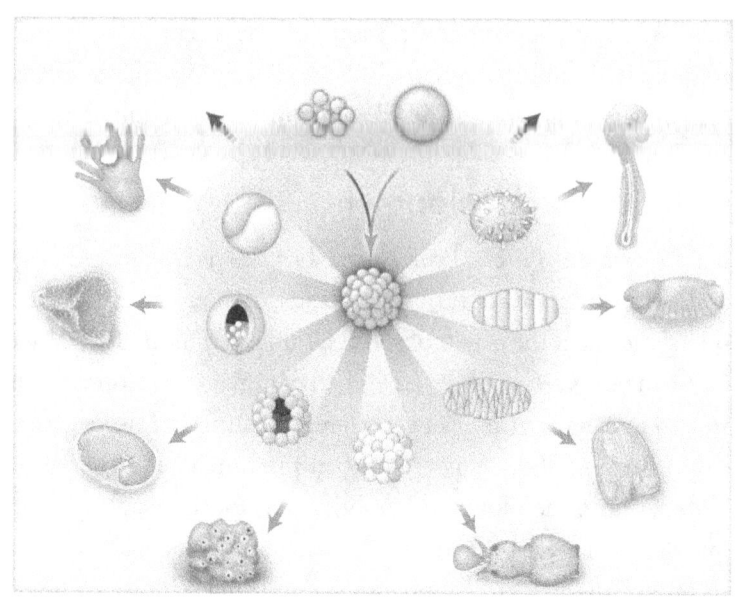

The inner circle shows some of the key emergent motifs (clockwise from top): appendages, segments, elongated bodies and primordial, coexisting alternative cell types, interior cavities, dispersed cells, and multiple layers. Contemporary organisms containing some or all of these motifs are shown in the outer circle (clockwise from top): vertebrate, arthropod, ctenophore, cephalopod, sponge, nematode, placozoan, echinoderm.
Source: Science, 12 October 2012: vol. 338, no. 6104 (12 Oct. 2012).

If his hypothesis is correct, numerous animal lineages arose and evolved independently of each other, yet share a patchwork of common physical and genetic elements depending on how the unicellular genes were repurposed for multicellular functions. He proposes that vertebrates, insects, nematodes, mollusks, and other major categories of animals had separate multicellular origins and leaves the obvious conclusion that each of those could have arisen multiple times.

Newman also claims that his model of "physics acting on early multicellular forms to define in broad strokes the patterns of development" resolves several incongruent ideas in the theory of evolution. For example, the rapid emergence of nearly all body plans in two short explosions of life, the use of the same genetic tool kit in disparate animal phyla, the recurrent appearance of the same animal body plans and organs, and the increasing complexity of animal body plans and organs (requiring new motifs) cannot easily be explained in the traditional Darwinian view of common ancestry. Newman leaves the lines of descent among major categories of animals as "uncertain".

Embryo geometry

Similar in many ways to Newman's work, a team from various universities spent 20 years trying to address the "foremost" unresolved problem confronting modern biology: to identify the origin of complex animal forms (aka, the origin of species).[116] Their new model, embryo geometry, offers "an intuitive and plausible description of the emergence of form via simple geometrical and mechanical forces and constraints".[117]

Their theory, published in 2016, reveals a different view of life than that which has prevailed for the previous 70 years. It proposes that the vertebrate body is mostly the product of mechanical forces and the laws of geometry, not random genetic mutation.

> We believe that the most plausible account
> of the emergence of the vertebrates begins
> with the self-organizing blastula: namely, the
> geometrically biased and mechanically constrained
> outcome of the subdivision of the zygote.[118]

The theory of embryo geometry challenges the idea that DNA contains the blueprint for the body. Rather, the body comes about based on the geometry of the blastula, the little ball of cells that grows into an animal. Musculoskeletal, cardiovascular, nervous, and reproductive systems all form through natural variations of geometric patterns. These universal forces mold both the individual embryo and the species over evolutionary time.

Newman's developmental motifs and embryo geometry both provide strong support for the idea of polygenesis; that creatures are the result of natural laws, that they will form repeatedly when conditions are propitious, and that common body forms are derived from geometry and physics. If an independent lineage is established when multicellularity is achieved, then the animal kingdom has numerous lines of independent descent, but the most significant advancement has been the emphasis on process and repeatability. Certainly, if these transitions are natural and follow repeatable pathways, we should expect multiple origins.

Polygenesis

The brain is possibly the most complex machine in the universe. Any life that does anything other than lounge around all day will have some kind of nervous system, even if it is decentralized like a brainless jellyfish. The creation of the first brain is currently believed by the generality of scientists to have only happened once in two billion years, stumbled upon during a blind walk through the limitless genetic space. The generalized common ancestor of all neuron users is a segmented worm in the early Cambrian.[119] This narrative is repeated throughout educational literature.

Recently, however, Dr Leonid Moroz of the University of Florida published a paper in 2009 firmly stating, "Molecular data and novel animal phylogeny imply that both complex brains and neurons evolved at least 5-7 times during the course of animal evolution."[120] Moroz refers to his hypothesis as "polygenesis", or "independent origins of neurons and complex brains among species in different lineages". He also says that "multiple origins better explains the extant diversity of nervous systems", rather than requiring derivation from a common ancestor for over 30 phyla (singular: phylum).

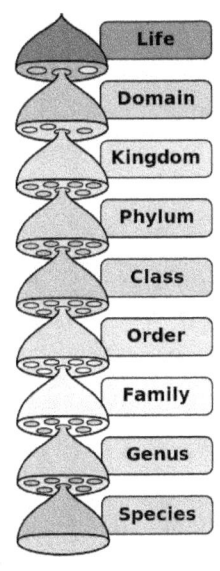

Moroz lists several animal phyla whose neural networks imply parallel and convergent development: vertebrates, mollusks, arthropods, nematodes, and annelids. He further writes that even these possibly have numerous origins within phyla. For example, mollusks include 3–4 cases where independent centralizations of their nervous systems might have occurred in parallel.

By breaking up the animal kingdom into at least five independent origins, and hypothesizing of "even more" origins within phyla, Moroz provides strong support for the independent descent of many species. Combined with Newman's developmental motifs and embryo geometry, there is a strong case that vertebrates have a unique origin separate from other animals, and that further research may identify several origins within vertebrates.

A unique origin of vertebrates is certainly in the direction of supporting parallel evolution, but the idea of all vertebrate life descending from a generalized common ancestor still would conflict with my understanding of 'Abdu'l-Baha. There would need to be many more vertebrate origins.

Polyphyly

There are not established unique origins within vertebrates, only speculation. There is, however, one notable source of speculation on repeatable evolution among vertebrates.

Malcolm Gordon, professor of evolutionary biology at UCLA, wrote a paper in 1999 with the subtitle 'A Speculative Essay'.[121] Speculative because Gordon suggests that two major evolutionary innovations should not be regarded as singular events: the origin of life, and the origin of vertebrate tetrapods (vertebrates higher than fish).

Gordon feels that evolutionary biologists have a near ideological commitment to the universal tree of life that has blinded them from evidence of multiple origins. In this context multiple origins is not meant lines of descent originating from a unique multicellular event, but evolution repeatedly arriving at the same optimal form that is misinterpreted as a singular event.

Most biologists start by assuming common ancestry and then try to place all organisms on the perceived tree of life. The placements end up being self-reinforcing. The paper suggests that single origin among the highest categories of taxa (kingdoms, phyla, classes) is "probably not" supported; that the "mosaic and chimeric structures of genomes" are factors discouraging single origin; that "important groups of organisms" probably had multiple origins; and that the "near universal search for monophyletic origins" is occasionally too narrow-minded and causes the omission of better models. (A monophyletic group is one that descended from a single common ancestor, a polyphyletic group is one that is not defined by common ancestry)

Gordon proposes that the universal tree "probably had many roots" and that the standard concept of common ancestry that appears at the taxa of species, genus, and family, is less and less supported while moving into the deeper categories of order, class, and phylum. He suggests that,

> [It is] nearly impossible that the highest categories of living organisms can be said to have had basal species... The traditional version of the theory of common descent apparently does not apply to kingdoms as presently recognized. It probably does not apply to many, if not all phyla, and possibly also not to many classes within the phyla.[122]

Gordon's speculative essay aligns very closely with my own thoughts. It is clear from Doolittle, Newman, and others that multiple phyla do not have an ancestral species that could be called the common ancestor. This is not just a case where the data fit both models; the data seem to only fit the model of multiple origins. Gordon recognizes the need for polyphyletic origins among phyla (arthropods, mollusks, vertebrates) and speculatively extends that idea into class (reptiles, birds, mammals), order (rodents, primates, carnivorous mammals), and maybe even family (dogs, cats).

These ideas are not in conflict with the theory of evolution, merely a refinement and clarification of the presumed branching pattern. However, the changes get closer and closer to a paradigm shift the farther up the tree of life that unique origins are identified. No eyes were batted when a unique eukaryotic origin was established for plants; it barely made the news. It is fairly uncontroversial to suggest that insects and vertebrates evolved on separate paths back to single cells; Newman's proposal resolves several incongruencies that are difficult to explain with a common ancestor. A separate origin for fish and reptiles would certainly be controversial. But what if all cats came from their own line of independent descent, separate from dogs? That would truly be a paradigm shift, and it would validate my understanding of 'Abdu'l-Baha's comments.

Independent birth and genomic potential

In 1994 Dr Periannan Senapathy, president of the biotech firm Genome International, proposed that the primordial pond of simple cellular life has the potential to create viable genomes and that evolution happens at the molecular level before the form appears at the multicellular level.[123] Thus, according to Senapathy, each species literally comes out of the genetic soup in a kind of stem cell that has a distinct mature form. When a new species forms in this manner, it then goes on to adapt into a variety of similar forms. So, for example, all snails might have a common origin, but their genealogical descent is original and non-overlapping with other organisms. The genetic and physical similarities between organisms, according to Senapathy, are due to their independent births from a common primordial pond, which is dominated by the horizontal exchange of genetic building blocks.

The striking justification that Senapathy provides to the model of independent descent is obvious, and Baha'is may be tempted to agree without taking a deeper look. Senapathy has gained little to no ground among his colleagues and makes the odd claim that evolution is not ongoing and that no new creatures will ever appear on earth.[124]

Independent of Senapathy, Dr Christian Schwabe of the Medical University of South Carolina published his own work in 2001, titled *The Genomic Potential Hypothesis*,[125] and a shorter article in 2008 titled 'Embryotic evolution'.[126] In some ways, Schwabe succeeded where Senapathy failed. Both of their conclusions similarly support the independent origin of thousands of species, but their paths differ on a few key points. Senapathy views species as forming out of a unicellular common gene pool, with an emphasis on horizontal gene transfer. Schwabe views species as forming at the crossing from inanimate to organic life, with an emphasis on convergence (see next chapter). Thus, Schwabe views prokaryotes as having an origin separate from eukaryotes and the mature form of a eukaryote species is not determined until it appears as macroscopic life. In other words, early evolution is highly repeatable and has a wide range of flexible outcomes, but the possible outcomes narrow considerably as a species develops towards complexity.

As a biochemist, Schwabe sees the idea of a single origin totally implausible and requiring several miracles. He favors multiple origins and inevitability in biological events, noting that the almost immediate molecular self-assembly that happened in the early earth indicates that it was no accident, and he predicts that "innumerable origins" evolved at many places on earth based on chemical affinity. Similarities, such as the universal genetic code, are based on basic principles of chemistry, not an accident frozen in time.

The genomic potential hypothesis proposes that a gradual accumulation of complexity happens in the primordial genome and this genomic evolution then spews out macroscopic organisms into the biosphere that appear increasingly complex, most of which are unsuccessful.

Schwabe sees no evidence in the fossil record that there was ever a single lonely species, nor solid evidence for the macroevolution that is postulated to make the most dramatic of transformations. Rather, the oldest fossils include diverse species, a fact that implies parallel development, and the only examples of speciation that are observable in a human lifetime involve minor variations to existing forms. For example, a species of fly can be separated into two groups, then bred for different traits over countless generations until they no longer interbreed, but they still look like flies. It is only speculative that these minor variations can lead to major variations given enough time, so that a salamander could evolve into a horse. The fossil record too, according to Schwabe, indicates that species appear first in very small embryonic forms, and then have no significant changes once a species is established in mature form.

The emphasis on natural laws and repeatability provides a contrast to both creationism and the prevalent model of evolution. It is also testable. Schwabe makes predictions about the genomic potential of complex life being found in simple forms. If these immature species are as prevalent as Schwabe suggests, they will be slowly discovered as genetic sequencing becomes faster and cheaper (see Chapter 9).

Senapathy's and Schwabe's theories would support a model of parallel evolution for many species. Schwabe's publication by a reputable source (Landes Bioscience) will interest many Baha'is. The question is, do these ideas herald a major paradigm shift in evolutionary theory? Maybe. Most ideas proposing a paradigm shift turn out to be wrong, and both of these theories seem to have faults. I believe a major paradigm shift is entirely plausible in this case, but I do not recommend a wholesale acceptance of either Senapathy or Schwabe.

Independent descent

The new view of independent starting points for genealogical trees coming from a tangled genetic web means that the assumption of *universal* common ancestry is incorrect as commonly interpreted. The tree model is a useful tool to describe complex relationships, but it is not applicable before the species threshold has been crossed, which is the point when a stable genealogical trace has been established and vertical evolution begins.

There are many questions to be answered still, but the trend is clearly supportive of a model of independent descent. Separate origin of many phyla is supported by many researchers asking the right questions, and further research may confirm several origins among vertebrates.

CHAPTER 5

CONVERGENCE

Convergent evolution

Common ancestry is a powerful explanation for similar traits among creatures. Why do all mammals have two eyes and not one or four? A popular understanding is that all mammals were derived from the same common ancestor, and it had two eyes, and therefore all mammals have two eyes. But there is another explanation for similar traits, one that is already widely known among evolutionary biologists: similar traits evolve independently in separate lineages because they provide to the bearer a selective advantage. This is known as convergent evolution, and the traits are deemed analogous. Convergence has been demonstrated in features such as bone structure or vision in disparate species where the proposed common ancestor did not have the trait.

For example, sharks, whales, and mosasaurs (a dinosaur that lived abundantly 90-65 mya) all share a similar body form that is optimal for hunting in water, and it obviously developed independently in fish, reptiles, and mammals.[127] Wings came about independently in birds, mammals, insects, and reptiles. The convergence of form suggests that an optimal economy is being found repeatedly in the evolutionary process and it amplifies our understanding of natural selection.

Convergence has been pointed to by several Baha'i authors probing the possibility of parallel evolution. Eberhard von Kitzing briefly mentions it in *Evolution and Baha'i Belief* (2001):

> Parallel evolution would be plausible if the space of possible forms of living would be strongly bounded and the transition within these possible forms along the developmental line of a species very likely. Such type of evolution is generally designated convergent evolution... To establish parallel evolution one would have to prove that due to bounds, within which life is possible, the reinvention of the same organs, of the same organelles, and often the same or very similar DNA sequences was inevitable. Without such a proof the model of parallel evolution would remain unsubstantiated.[128]

For the model of independent descent to be valid, convergence would need to play a far greater role than imagined in mainstream science. The idea persists that the formation of numerous biological novelties are singular, without parallels over time. The emergence of animal forms would have to be shown repeatable and, in a sense, inevitable. If these biological novelties do not follow constrained, repeatable pathways, then indeed the universal tree of life can be constructed by the principle that similarities imply closer ancestry, and life of independent origin would spawn radically different forms.

This has to do with what is probable, but how would such a thought experiment be validated?

Replaying the tape of life

One of the most intractable questions to biologists has been whether the outcomes of evolution are incredibly variable and based on chance contingencies, or limited and repeatable. What are the roles of contingency and determinism?

Stephen Gould poignantly (and brilliantly) asked the question in *Wonderful Life* (1989): if the tape of the universe were rewound to before the Cambrian explosion and allowed to play again, what would the result be? Visible life before the Cambrian (>560 mya) was limited to jellyfish, worms, and a few very simple plants. Gould suggested that the results would be radically different, not producing anything like *Homo sapiens*. He was forthright in claiming that human intelligence has a meaninglessly small probability in the course of evolution.

In the following decade, many scientists began testing the repeatability of evolution, and finding that the outcomes of natural selection are not so random after all.

An interesting new star in the world of evolutionary biology is Simon Conway Morris, a Fellow of the Royal Society who earned notoriety in his work on the Burgess Shale fossils and the Cambrian explosion, and who was described by *The Economist* as "the champion of a new interpretation of evolution".[129] As a direct response to Gould's question, he published *The Crucible of Creation* (1998) as an attack on the contingency emphasized by Gould. With a different interpretation of the Cambrian fossils along with examples of convergence and constant evolutionary pressure, he argues that in the broadest terms evolution is predetermined and will always follow a similar path.

Conway Morris published further on the subject of convergent evolution in *Life's Solution* (2003) and *Runes of Evolution* (2015). In these books he continues to document pervasive examples of convergent evolution, indicating that the outcomes of evolution on earth are far more limited than previously imagined. What is biologically possible, he writes, "has usually been arrived at multiple times, meaning that the emergence of the various biological properties is effectively inevitable". But his main purpose was not to act as a compilation of examples of convergence, but to argue that, "contrary to received wisdom, the emergence of human intelligence is a near-inevitability".[130]

Pervasive convergence

Jonathan Losos, professor of biology at UC Berkley, was just finishing grad school when Gould's *Wonderful Life* appeared in 1989. He read it voraciously and was convinced of its argument that the outcomes of evolution are so variable that if the tape of life were allowed to replay, the outcome would be totally different. In Losos' own book in 2017, he says that Gould started a quarter century of debate over the repeatability of evolution, but was effectively countered as Conway Morris and others articulated the implications of convergence. He summarizes:

> An intellectual counterpoint to Gould's emphasis on unpredictability and non-repeatability has emerged. This alternative view emphasizes the ubiquity of adaptive convergent evolution: species living in similar environments will evolve similar features as adaptations to the shared natural selection pressures they experience... convergence demonstrates that evolution, far from being quirky and indeterminate, is actually quite predictable: there are limited ways to make a living in the natural world, so natural selection drives the evolution of the same features time and time again.[131]

> The standard wisdom is that convergent evolution happens, but is not necessarily the expectation... But all that is changing. In recent years, a cadre of scientists has taken the opposite view, arguing that convergence *is* the expectation, that it is pervasive, and that we should not be surprised to discover that multiple species, often distantly related, have evolved the same feature to adapt to similar environmental circumstances. From this perceived ubiquity, the scientists draw a broader conclusion: evolution is deterministic... the contingencies of history play a minor role, their effects erased by the predictable push of natural selection.[132]

The role of convergence has been rising in mainstream scientific thought over the course of 25 years, and there are many such authors similarly suggesting that the evidence points to prolific convergence, with all its implications. I'd like to share one more excerpt published in 2006 to emphasize the breadth of sources and forceful language used by the "cadre of scientists" that are writing about those implications.

Geerat Vermeij of UC Davis wrote:

> [If] even very rare phenomena can be shown to be iterative and replicable, and if certain pathways and outcomes are strongly favored over others, then similar phenotypes and interactions of life should emerge wherever conditions suitable for life exist. History therefore would be predictable at the scale of phenotypes, ecological roles, and directions of change, but it would be contingent in the details of initial conditions, pathways, players, and timing...

A literal reading of the history of life implies that many events, including evolutionary breakthroughs, occurred only once. Data and arguments from various sources, however, indicate that few, if any, innovations are truly unique. Indeed, the principles of physics and economics imply that many derived functional states are achieved many times in many clades because they impart substantial, widely applicable advantages to their bearers...

But there is strong evidence from evolutionary convergences that the transitions are not random. Some configurations stabilize and self-organize more readily than others, and economic selection strongly favors some directions and some functional outcomes over others. These physical and economic realities therefore impart to history a certain predictability and replicability.[133]

During and after Conway Morris made his contributions, the popular press, which likes to describe convergence as 'shocking', picked up on the story. For example, in 2003 this New York Times article described more of this type of research. There are many like it.

In 1988, Dr. Lenski and his colleagues set up a dozen genetically identical populations of E. coli bacteria in bottles of broth and have followed their evolutionary fates.

Now, more than 30,000 bacterial generations later, Dr. Lenski and colleagues have what is becoming one of the most striking examples of repeatability yet. All 12 populations show the same patterns of improvement in their ability to compete in a bottle and increase in cell size. All 12 have also lost their ability to break down and use a sugar, called ribose.

More surprising, many genetic changes underlying these adaptations are very similar. Every population, for example, lost its ability to break down ribose by losing a long stretch of DNA from the same gene.

Other scientists studying cichlid fish have observed how the same varieties of cichlids evolve anew every time they invade a new lake. And Dr. Rieseberg and colleagues have found evidence that evolution can repeatedly produce the same species.

These scientists found that one sunflower species on sand dunes has evolved independently three separate times. And each time one of the species newly evolves, genetically it appears to turn out much the same. "With these species, there seems to be only one way to do it," Dr. Rieseberg said.[134]

Convergent traits

Life's Solution (2003) and *Runes of Evolution* (2015) by Simon Conway Morris, *Convergent Evolution* (2011) by George McGhee, and *Improbable Destinies* (2017) by Jonathan Losos provide numerous examples of convergence in biological evolution. Below is a small sample.

- Coffee, cacao, and tea all independently evolved the production of caffeine.
- Your eye is nearly identical to that of an octopus, though they are wired a bit differently. In fact, the camera eye has

developed independently at least six times, and by some counts at least forty times.

- The praying forearms used for capturing prey in mantises and mantidflies are nearly identical, as well as their necks and heads.
- The colonies of ants and termites are amazingly similar, down to the harvesting of fungus and antibiotics, though not derived from common ancestry.
- Your intimate friend, the tapeworm, has a form and life-cycle that parallels the Haplozoon, an unrelated parasite that shares its intimacy with marine worms.
- Euphorbs and cacti have both developed succulent traits and spines, and though the traits were not shared in a common ancestor, they are easily confused for each other.
- Giving birth to live young came about hundreds of times in everything from fish to amphibians to sea stars to insects.
- Three types of plants – the corpse lily, the carrion flower, and the Zulu giant carrion plant – all evolved to produce the smell of rotting meat to fool prospective pollinators.
- Powerful claws and a long sticky tongue, ideal for eating ants, evolved at least 10 times, maybe up to 30 times.
- The platypus injects venom from a spur on its heel, has a beaver tail, three sex chromosomes, webbed feet, and a duck-like bill that can sense electric movement for hunting in water – similar to sharks. The consensus is that all its traits, other than egg laying and mammary glands, are not derived from common ancestry.

The list goes on: molar teeth, eardrums, porcupine quills, fingerprints, opposable thumbs, beaks, sonar, voice, venom, toxic defense, mimicry, hinged shells, gyroscopes, electric organs, antifreeze protein, infrared vision, bioluminescence, and even such fundamental attributes as warm-bloodedness, placentas, four-chambered hearts, and brains. According to Conway Morris, any trait that you can think of has come about multiple times.

Convergent forms

According to theory, the common ancestor of marsupial and placental mammals was a rodent-like creature that produced the first mammary glands. Later, one of its rodent-like progeny made an evolutionary advancement by making a better use of its placenta. Any novel evolutionary traits in disparate marsupial and placental mammals should be the result of convergent evolution, not common ancestry. The following animals from the two lineages fill similar niches and have such similar bone structures that they are sometimes used as a test to trick biology students:

- The marsupial mole and the mole
- The marsupial mouse and the mouse
- The marsupial lion and the lion
- The marsupial sabre-tooth and the sabre-toothed cats
- The rabbit-eared bandicoot and the rabbit
- The sugar glider and the flying squirrel
- The wombat and the groundhog
- The quoll and the wild cat
- The thylacine and the wolf

Outside of the marsupial/placental examples, similar body forms have been produced independently all over the world. The peccary and the wild boar (see illustration next page), the rabbit and the mara, the kangaroo rat and the hopping mouse all share body forms that are not derived from common ancestry. The list goes on and on.

Nearly all the examples above have been identified as convergent because the features and forms developed in unrelated species, in cases where the proposed common ancestor did not have the trait. The similarities imply that there is an optimal solution that is being sought out in the evolutionary process to suit basic needs. The differences imply that the solution evolved more than once.

The wild boar (top) and peccary (bottom). Only related by marriage.

Convergences are interesting and show the repeatability of natural selection. Many leading experts agree that the pervasiveness of convergent traits should change the traditional view of a haphazard process creating unlikely outcomes.

Although it shows a shift in the direction of agreement, this view does not by itself validate the independent descent of humans. Most of the examples given are slightly different in design. They might just perform the same function using different methods. Convergent evolution isn't a separate kind of evolution and doesn't conflict with the idea of common ancestry for all vertebrates. If many of today's species come from independent descent, even with many similarities from a common gene pool, convergence will have to arrive at nearly identical forms, so close that science has been confusing them for over a century.

The flat fish

Turbot, a species of flatfish.

Darwin's view of speciation was admittedly confounded by the sudden creation of new characteristics, and the lack of intermediary transitional fossils. For example, flatfish such as halibut are one of few animals without external symmetry. During life one eye migrates to the other side, leaving the fish flat and giving it a selective advantage in bottom-feeding or in shallow water. Darwin was puzzled by flatfish and thought there must be fossils showing intermediary forms, but none could be found and an intermediary form would have no selective advantage, so he had to leave open the possibility of a sudden change from one generation to the next. In 2008, a transitional fossil was found with a flatfish's lower eye mostly, but not entirely on the other side of the head, thus showing gradual transition.[135] In the debate over the flatfish, the point was missed that some lie flat on the left side and some lie flat on the right side. How many times did a flat fish evolve?

In the case of flatfish, the same selective pressure and environment caused the same result. Boned fish naturally develop like a pancake standing on edge, but as some fish went to bottom-feeding, their body shape made one of their eyes useless, as it was in the sand while they fed horizontally. Normally a mutant fish with both eyes on the same side of the head would be less fit and fail to reproduce, but for bottom-feeders the mutation provided an advantage, so the mutation was propagated into an asymmetrical oddity.

Some flatfish species developed on the right side and others on the left, and there is no selective advantage to left over right. The transition from vertical fish to horizontal fish, and a corresponding helpful mutation of the eye, happened more than once. There are two ways to do it, and we see both in nature, suggesting that it was not a unique transition. The trait of flatness in fish was achieved convergently in several species evolving in parallel, not derived from a common ancestor that diverged into all known flatfish.[136]

Asymmetry

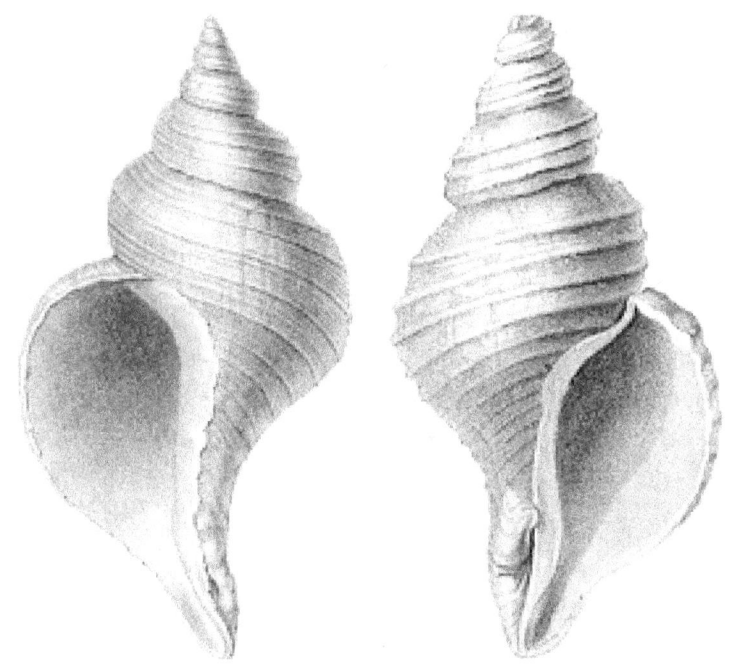

Sinistral (left) and dextral (right) shells.

Snail shells have to spin from a point in a binary fashion, either sinistral (left-handed) or dextral (right-handed). Similar to flatfish, within a species of snail, the entire population will be either left or right, and throughout the globe populations are found with differing orientations of spin, indicating independent origin.

Also similar to flatfish, within a species of left-leaning shells, an occasional right-winged snail appears, and the reverse is true of dextral snails. One might surmise that this occasional flop could result in a snail whose progeny form a new species. Thus, the attributes of flatness in fish and coiled shells in snails would not be clearly convergent. However, studies into the outliers have shown that they are a result of congenital irregularity, not genetic mutation, so the trait of shell reversal would not be carried on into progeny.[137] To further emphasize the point of convergence, there are snail eating snakes (*Pareatinae*) that have themselves developed more teeth on the right mandible in order to prey on sinistral snails.[138] The same snake would more frequently drop and fail to eat a dextral snail. If the spin direction of the shell were easily changed through mutant variation, the new form would be naturally selected until the predators can adapt. The direction-reversal of snail populations has not been observed, and the asymmetry of predator snakes has been found in both dextral and sinistral forms, depending on the orientation of prey. So despite the occasional reversals in a given population, there is a strong case for convergence of fish flatness and shell spin.

An exceptional example of asymmetry is found in the owl. While hunting at night owls rely heavily on sound. Imagine a barn owl perched on a branch when it hears a tasty mouse rustling in the grass. The intensity of the sound is greater in the left ear than in the right, and the brain also recognizes a time delay in one ear, so it turns its head to the left and narrows in on the direction. As it flies, the sound from the target creates a three-dimensional impression because the barn owl has asymmetrical ears, the left ear is high on the skull, and the right ear is low. As the intensity of the sound of the rustling mouse grows in the lower right ear, the barn owl begins to narrow in on the tasty treat, and thanks to the asymmetry of the ears, it easily nabs the rodent in total darkness and brings it back to feed its asymmetrical owl chicks. The owl is another case where asymmetry provides an advantage, but there is no distinguishable selective advantage to right or left. The result is that some species of owl have the left ear higher, others have the right ear higher, others have asymmetrical skulls, and yet others form asymmetry through soft tissue canals instead of skull structure. The various ways to achieve asymmetrical hearing led researchers to conclude that the technology was developed independently on at least five distinct genera of owl, "probably" more.[139]

The asymmetrical hearing found in owls is also found in toothed whales (*Odontocetes*) in both left and right form.[140] So the trait evolved not just several times among owls, but also several times among toothed whales.

The toothed whales also share the ultrasonic echolocation that bats use, and the proposed common ancestor of bats and whales did not use echolocation. Not only do bats and whales both use sonar, but they use the same range of frequencies (15-200 kHz) and both increase their click rates to about 500 per second while hunting.[141] The genetic adaptations for echolocation in bats and toothed whales are driven by nearly identical genetic changes.[142] So genes, too, are similar in totally unrelated animals due to convergence. Furthermore, an insect in New Zealand – the weta – and toothed whales both use a particular fatty tissue for hearing.[143] So in just the example of toothed whales convergence can be seen of asymmetrical hearing (both left and right form), fatty tissue for hearing, and ultrasonic echolocation.

The examples of asymmetry have an odd characteristic. They prove convergence not by developing the trait in totally separate lineages, but because there is a binary option on how to develop the trait, and neither option has a selective advantage over the other. The convergence seen in asymmetrical forms shows that natural selection produces the exact same results repeatedly. If the few examples of asymmetry have been produced repeatedly, then shouldn't we expect symmetrical characteristics to also have been produced repeatedly without our noticing? When the results are symmetrical it becomes incredibly difficult to distinguish them as homologous (similarities derived from a common ancestor) or analogous (similarities evolved independently).

Organ asymmetry

There exists one form of asymmetry in humans that presents a challenge to the theory of independent origins among mammals. The human heart has four chambers, one of which pumps blood to the entire body and grows much larger than the others. In humans the heart is larger on the left, not to mention that the right lung is larger, the stomach and spleen sit on the left side, and the liver has a lobe on the right side. This asymmetry is the same in all mammals.[144] Similar to flatness and shell rotation, in very rare cases a congenital abnormality (*situs inversus*) causes a complete mirror-image reversal of all asymmetrical organs. Although it is an abnormality, it is not a disability, suggesting that there is no advantage to one form over the other. The logic surrounding asymmetry would suggest that the mammalian body form appeared only one time in a generalized common ancestor.

However, the mechanism behind symmetry has eluded researchers and *universal* asymmetry raises many questions. A 2005 review of left-right asymmetry documented various attempts to discover the cellular mechanism behind it, but emphasized several open questions and the need for more research.[145]

An earlier model proposed that left-right asymmetry must be due to the chiral asymmetry of molecules being transferred to cellular and global asymmetry.[146] The common mammal asymmetry might be derived not from a common ancestor, but from a universal biological principle.

Since then many biologists have focused their research on the drivers of left-right asymmetry, and they have largely confirmed the theory that asymmetry is derived from cellular chirality. Research published in 2012 showed that the proteins involved in forming asymmetry are the same in all plants and animals. So if the trait comes from common ancestry, it originated long before multicellularity was a reality.[147] Then in 2016 there was another confirmation that "animals with diverse body plans may derive their asymmetries from the same intracellular chiral elements".[148]

As strange as it seems, the common organ asymmetry found in all mammals may not be strong proof of common ancestry. There may only be one way to do it, in which case the data can support both common ancestry and independent descent.

Optimal solution

On the outside, at least, humans are symmetrical. Their right arm mirrors their left, so that whatever appears on one side of their midline also appears on the other in the same general color and shape. From an evolutionary tree point of view, symmetry makes it more difficult to identify traits as either derived from common ancestry or independently evolved. When there is an optimal solution that provides an advantage, then it will be selected naturally, and the exact same trait may be selected in a similar species without leaving any definite signs of independent origin. As most features are symmetrical and optimal, there remains a dearth of evidence for convergence and a general bias towards assuming that common features derive from common ancestry.

This bias is also expressed looking back in time. Examples of known biological convergence become less frequent in the distant past, implying that either the early phases of life's history were less replicable than later phases or that results are biased from a lack of data on ancient creatures. This means that parallel development should be far more pervasive than imagined.

Are not the ideas of convergence and optimal solutions what 'Abdu'l-Baha recalled while disputing that "species evolve and can even change and transform into other species"[149]? He said,

... the underlying wisdom and cause of the differences
in the colouration of animals and of human hair, or
of the redness of the lips, or of the variety of the
colours of birds, are still unknown and remain hidden
and concealed. But it has been discovered that
the blackness of the pupil of the eye is due to its
absorbing the rays of the sun, for if it were of another
colour—say, uniformly white—it would not absorb
these rays. Now, so long as the wisdom underlying
the things that we have mentioned is unknown, one
may well imagine that the reason and wisdom of
the vestigial limbs, whether in the animal or in man,
is also unknown. Such an underlying wisdom of
course exists, even though it may not be known.[150]

The European philosophers viewed the lack of purpose in vestigial organs as evidence of two revolutionary ideas: species evolve and they share common ancestry. That species evolve is of no dispute, but as soon as function and purpose can be attributed to a shared feature, convergent evolution can explain the similarities without requiring inheritance from a common ancestor. Variation without a selection process would not converge on similar forms, but an attribute that adds to fitness will be repeatedly selected independently in unrelated organisms.

There is an advantage to forming two camera eyes, instead of one or three or seven. A single eye cannot perceive depth, and three eyes don't give an advantage comparable to the biological cost of the extra eye, so vertebrates almost always have two. There is also the question of bilateral asymmetry, which would probably deter an odd number of eyeballs, ears, or limbs.

With few exceptions, a compound pixelated eye is the most efficient for the size and use of arthropods. There is a simple competitive economy being carried out in the evolutionary process that naturally selects the greatest efficiency, and the results are repeatable.

A similar argument could be described for the general body layout common to all vertebrates regarding the placement of the head, neck, legs, organs, even the body temperature of mammals. The pervasive body plan found among vertebrates has been used as proof of common ancestry, but what other useful form exists while evolving through various stages of evolution? A big cube-creature with three limbs and one eye would quickly go extinct on the African savannah.

Four-limbed vertebrates have an incredibly efficient body plan. For all mammals, amphibians and most reptiles, having four limbs, whether a combination of wings, legs, or arms, is simply the optimal solution during the course of evolution. More limbs would not be worth the biological cost, less might imperil the creature to extinction, or other options might not have an evolutionary path. The elephant effectively turned its nose into an extra limb, but there are few evolutionary pathways to get there, and the elephant's need for an extra limb comes from its feet being purposed for being enormous instead of fine manipulation.

Not every living organism is perfect. Less optimal forms still appear because there may not be an evolutionary path from here to there. The mythological centaur would have to evolve from a six-limbed primate or horse, which are inferior forms that would go extinct when competing with the four-limbed varieties, thus no centaurs evolved. The flatfish is an obvious example of a less-than-optimal design that came about from limited evolutionary paths. The same goes for the human style of birthing large-headed offspring.

Better than a human?

Few ways to do it

It appears the principle of convergent evolution is more radical than at first glance. Genealogical trees have long been designed around the principle that similar physical and genetic traits are almost always derived by inheritance from a common ancestor. We tend to picture evolution as a series of singular evolutionary events. When an unlikely novelty arises, it comes about once and evolution of that novelty simply stops.

In traditional evolutionary studies, traits are only confirmed as convergent when no other possible explanation exists. But evidence has mounted to change this paradigm. One of the many implications of prolific convergence is that no biological trait on earth is truly novel. Another is that evolution of traits and forms should be viewed as repeatable processes, not events.

Now there are three tools in the biologist's kit to explain similarities, and each of them appears to be an equally powerful explanation: inheritance from common ancestry, shared descent from a common root system, and convergence.

Convergence goes beyond shared body forms but includes gene networks. Surprisingly, the principle of convergent evolution applies to the deletion,[151] addition,[152] and substitution[153] of genomic data in complex organisms, and the universality of the genetic code.[154] This is because, contrary to popular belief, "environment initiated novelties" have a greater effect on evolution than gene mutation, meaning that "genes are probably more often followers than leaders in evolutionary change".[155] If the environment shapes gene networks, then convergence should be expected in genes (see Chapter 9).

Inevitable pyramids

An example can also be drawn from social and cultural evolution. When the same cultural attribute appears in disconnected parts of the world, does it result from a common origin? A series of persuasive History Channel documentaries called *Ancient Aliens* propagated the idea that advanced alien races visited earth in the past and that this contact was the origin of human technology, religion, or biology. The idea of ancient aliens is often supported by the complexity of the Egyptian pyramids, and similarity to pyramids in Mexico, Cambodia, and China at a time when those cultures had no known contact. A long list of other coincidences in cultures all over the world was presented.

Homo sapiens in Mesoamerica and Mesopotamia developed

similar levels of culture at about the same time following the last ice age, with no known communication between them. To develop such a society, each used language, developed tools, formed a government, and produced a calendar. Each had a religion and a belief in the afterlife, and within each of their languages, a word formed to describe justice, with the corresponding role of judges.

At a certain point this combination of social reality produced the desire to build something really tall that would last a long time, and with the resources available at their level of development, they had to use big blocks of rock. When building with blocks, there is an optimal way to stack so that they will be really tall and the structure will last a long time. There are universal rules of gravity, geometry, and structure. If the blocks were stacked in a tower, they would fall over. What option is there other than a pyramid? The most stable and tall structure would be a cone at about 45 degrees from the ground, but a cone cannot be efficiently built with stones, so a pyramid is the result. Two cultures of similar technology independently converged on the same optimal solution to the same need, and there is no reason to invoke alien helpers to explain their existence. Their similarities are not due to common ancestry.

Counterexamples

There are also many examples against prolific convergence in nature. Why didn't evolution produce another aye-aye of Madagascar, or platypus of Australia? What about kangaroos, kiwis, or giraffes? Why did we only see the early Cambrian animals once? Sharks and whales are similar, but sharks have vertical tails and whales have horizontal ones. Penguins exist on the south pole, but not the north pole. All the strange traits of the platypus must have been evolved convergently in other species, but evolution has not produced a counterpart to the platypus. All evolutionary outcomes are clearly not inevitable.

Obviously, the reality of evolution is complicated. Historical and environmental contingencies do play a role. Already-filled niches might preclude the pathways necessary to evolve a similar species. As competition grows, less optimal pathways stop producing viable species. Creatures are shaped by the environment into similar patterns, but they don't all face the same environments. Evolution does not produce perfect results, it creates embellishments on existing basic forms and needs the intermediate steps to create complexity.

Jonathan Losos suggests that we should view convergence as prolific among similar creatures, suggesting that the outcomes of evolution are highly constrained by starting points and options along the path of evolution. He says,

> ...species may end up stuck with suboptimal adaptations. For whatever reason, their ancestors didn't embark on the best road to adaptation. Natural selection pushes the species along, and it ended up adapted, but not as well as it might have been. This reasoning emphasizes the role that contingency may play in determining evolutionary direction and why, as a result, species may fail to converge when faced with identical environmental conditions...
>
> By the same logic, we might expect that the more similar two ancestral species are, the more likely they would be to evolve in the same way when facing similar selective conditions. And that is exactly what happens.[156]

Populations of similar species often repeatedly make the exact same transitions, physically and genetically, especially when faced with similar environments and opportunities. For example, a bird species (*Dryolimnas cuvieri*) once colonized an Indian Ocean atoll and became flightless. The island became inundated with rising oceans and the species went extinct. About 20,000 years later the exact same species appeared again because the same flying ancestral species was still nearby colonizing islands and again became flightless.[157] In this case the same progenitor species and the same opportunity created the same transition. On a grander scale, 99% of all bird species are combinations of the same 2-4 feeding niches, indicating strong convergence of forms, but we're interested in the other 1%, the one-offs, and whether the transition into their species was a species/opportunity combination that is truly unique or improbable.

In Conway Morris' third book, *Runes of Evolution* (2015), he responds to critics that point to evolutionary one-offs like the woodpecker and aye aye. The woodpecker has adapted to bang its head against a tree as a percussive instrument to find grub holes, then it opens the hole and uses a long tongue to nab its grubby victim. The aye aye is a primate that uses a long bony middle finger to percuss the tree while listening with sensitive ears. When the aye aye hears a grub hole, it uses teeth to open the hole, then the percussive finger of death nabs the grub. If evolution is deterministic, why are there such variable ways to get grubs? Where is the marsupial aye aye? If the woodpecker is unique, why can't humans be viewed as an unlikely evolutionary outcome?

Along with many other complaints against his philosophical views, Conway Morris responds with a deluge of examples of convergence. Modern woodpeckers appear to be descended from a single head-banging ancestral species, but among the birds of Madagascar, one species (*Falculea*) plays the woodpecker role of rapid percussive impacts, as does the 'akiapola'au of Hawaii, and the *Glyphorynchus*, and the ovenbird Xenops. The aye aye, then is also viewed as a remarkable example of convergence, providing a variation on the percussive hunter and showing how the role can be easily filled from different starting points. But isn't the aye aye an evolutionary one-off? No, says Conway Morris, percussive foraging is "a behavior that finds a convergence in the capuchin monkeys of the New World".

> This arrangement is evidently too good an invention to use only once, and striking parallels also exist with the marsupial possum *Dactylopsila*. Peering out of the fossil record are three more examples of honorary woodpeckers. Best known are the extinct Eocene apatemids, but more recently strong cases have been made for examples from the Miocene, especially a notoungulate (*Hegetotherium*) from South America and a marsupial (*Yalkparidon*) from Australia. Robin Beck aptly refers to this example of "a mammalian woodpecker... at least one of its manual digits may be elongate".[158]

Conway Morris views evolutionary determinism not in specific outcomes, but in patterns that he calls "ecomorphs – recurrent anatomical configurations that answer the call of particular ecological needs."[159]

Similar stories can be recounted for most of the one-offs. What about the lack of penguins in the north? Well, go back 25 to 34 million years and you'll find penguin doppelgangers swimming in the arctic.[160]

In a non-biological example, consider the form and function of an airplane. The creation of thrust to fly through the air can be created many ways, but the possibilities are limited, and so are the generic forms that efficiently fly. The airplane was contingent on a variety of other technological breakthroughs that did not have the airplane in mind, so the specifics of an airplane design are variable, but the ability to fly through the air would have been assembled eventually because it provides substantial benefit. Without Michael Faraday or the Wright brothers would we be flying today? Certainly. Rewind the clock and allow chance contingencies to play out. You won't see another Boeing 737, but you will see something that shares many similarities with it and fills the same role.

Similarly, evolutionary determinism means that the opportunity to consume nutritious grubs created strong selective pressure to exploit them with percussion. Given the numerous examples of percussive foraging for grubs, that evolutionary outcome is effectively inevitable.

Divergence vs convergence

In *On the Origin of Species,* Darwin actually responded to a criticism that he ignored convergence in his theory.

> Mr. H.C. Watson thinks that I have overrated the importance of divergence of character (in which, however, he apparently believes), and that convergence, as it may be called, has likewise played a part... But it would in most cases be extremely rash to attribute to convergence a close and general similarity of structure in the modified descendants of widely distinct forms... It is incredible that the descendants of two organisms, which had originally differed in a marked manner, should ever afterwards converge so closely as to lead to a near approach to identity throughout their whole organisation.[161]

Darwin didn't totally dismiss convergence. He said that natural selection "sometimes" resulted in similar parts, owing little to common inheritance, but he felt that convergences would be too incredible to be true for the whole species. As we see in this chapter, Darwin is incorrect in spirit, but wins on a technicality. He defines convergence as organisms that "originally differed in a marked manner" coming to very similar results "throughout their whole organisation". Evolution doesn't converge from wildly different starting points, but it does produce similar results from similar starting points.

Where generations of scientists have faltered was to designate gradual divergence of character as the overarching principle of evolution. There are two ways to look at the same data. If the beaks of finches adapt over many generations to their food source, naturally selecting either short fat beaks or long narrow beaks, then the same selective pressure will change the short beaks back into long beaks when the food source changes. The same selective pressure would also form similar beaks in unrelated species. This should be rightly regarded as a kind of convergence.

However, in the case of divergence, the small-scale examples have been assumed to be the rule to explain the macroevolution of large-scale changes, even though macroevolution is hardly observable. Repeatability and convergence has never been given such a generous presumption, even though the established small-scale examples of convergence are prolific.

In fact an extensive study of the Galapagos finches between 1973 and 2003 found that,

A combination of hybridization and selection jointly provide the best explanation of convergence in morphology and genetic constitution under the changed conditions following a major El Niño event in 1983. The study illustrates how species can alternate between convergence and divergence when environmental conditions oscillate.[162]

Even Darwin's finches demonstrate both divergence and convergence.

Evolutionary constraint

The lesson of convergent evolution is about evolutionary patterns, processes, pathways, or predispositions. Some people also call these evolutionary constraints. As a species grows in complexity, it becomes resistant to change, or constrained. There is not a pathway from scorpion to mammoth, but there may be a pathway from pig to mammoth, and once that pathway is known, it could have happened numerous times when the same opportunities are presented to a similar ancestral species. The mammoth, however, seems more constrained in what it can evolve into. As Carl Woese put it for cellular evolution,

[T]he more complex, integrated, and specific a cell design becomes, the more intolerant of change that design is.[163]

'Abdu'l-Baha confirms the idea that species change and evolve within a constrained range.

… in the vegetable kingdom… we observe that the original and distinctive character of the species does not change, while its form, colour, and mass do change, transform, and evolve.[164]

Combining these repeatable, evolutionarily constrained path ways with the new generation of multicellular forms, one can see a model forming where a new species is generated in one of a limited set of generic forms, creating a repeatable motif. That motif begins to embellish on the generic forms, but from the beginning it has certain constraints and eventually goes down the path of predator or prey (or both) and chooses speed vs camouflage. Every embellishment both precludes some options and opens up others. As Geerat Vermeij put it,

> History therefore would be predictable at the scale of phenotypes, ecological roles, and directions of change, but it would be contingent in the details of initial conditions, pathways, players, and timing.[165]

We are starting to see the outline of a model of many species forming and evolving independent of each other within constrained pathways.

CHAPTER 6

RAPID EVOLUTION

Guppies in streams

Traits have long been labeled as convergent when they appear in distantly related organisms. They are also labeled as convergent when they appear in asymmetrical form. Another method of identifying convergence comes from geographic isolation.

For example, cichlid fish have adapted to freshwater lakes around the world, and each isolated population made almost exactly the same adaptations over and over. Fish have also adapted to isolated cave systems all over the world and repeatedly and independently lost their sight and pigmentation.[166] On islands all over the world, elephants have miniaturized, birds have lost their wings, and shrubs have grown into trees.

Evolutionary biologists for over a century did not experiment with evolution because they viewed it as a slow process over thousands or millions of years. It turns out a century of biologists were wrong about the speed of evolution.

Evidence began with the speckled moths of England that turned dark while the industrial revolution darkened the tree barks where they sat. People took notice when antibiotic and herbicide resistance started appearing in the 1950s. A detailed look at evolution started in 1973 by studying the Galapagos finches. Then in 1980 a bombshell paper was produced by John Endler that began a new era of field experiments in evolution.

The island of Trinidad has many isolated streams rolling down from mountains. The streams tend to have several pools of water flowing over waterfalls of varying heights. Mr. Caryl Parker Haskins previously did extensive field work on the island showing that the wild guppies that inhabit the streams vary in color as they move up in elevation. The tiny fish are brown closer to the ocean, and more colorful closer to the mountains. His work also indicated that this coloration is due to the presence or absence of predators: the killifish and pike cichlid. Those higher pools that lacked a pike or had only the less-lethal killifish had the most colorful guppies, and in lower streams with more predators the guppies were a drab brown. The colorful spots on the flamboyant male guppies were very effective at attracting females, but also attracted bad guys. The repeated results in independent streams and pools showed convergent evolution.

Along comes John Endler, who decided to test the theory that the predator fish is the reason for the changes in evolution, as opposed to some other factor like bird predation. He used a greenhouse in New Jersey to re-create the streams of Trinidad, including pools and waterfalls. He took a large population of guppies from several wild streams and allowed them to interbreed for awhile before putting 200 in each of the streams. After four weeks in their new habitats, he dropped a highly predatory pike cichlid into four pools, a less-predatory killifish into four pools, and left two pools without predators. Other than daily feeding and water monitoring, the experiment was left on its own.

Evolution was supposed to take a long time, and guppies can't reproduce until about two months old, so he decided to check the experiment at five months. He initially measured every fish's size and counted their spots, when they were a mix of ornamentation. To his surprise, at five months they had already diverged in colors and spots. The pools with pikes had a ten percent reduction in spots, and the pools with no predator increased their attractiveness by ten percent. The guppies with killifish were somewhere in the middle but closer to the pool without predators.

Nine months later the difference was even more pronounced. Two years from the start of the experiment, the populations resembled the colors found in the wild while moving from the ocean to the mountain up a Trinidadian stream. The pikes were eating brown male guppies, and the free-loving male guppies with no predator were sporting large colorful spots to attract the ladies.

As surprising as this was, Endler knew that he was running an experiment in the lab, and it lacked the exact environment in Trinidad. Endler had also been visiting pools in Trinidad and identified some with only killifish. Killifish and guppies can bypass waterfalls by wiggling around them in shallow water during rains. Some pools had guppies without killifish, and others had killifish without guppies. Others had pike cichlids and killifish.

Endler visited one pool for over two years to make sure that it had no guppies and only killifish. Then he dropped in 200 brown guppies from a stream with pike cichlids. He returned two years later to find the previously guppy-less pool now had colorful guppies. He repeated his test in nature and it matched his lab results.

Others conducted several more similar experiments on Trinidad and got the same results, including adding a pike to a guppy-only pool and watching them turn brown with shorter lifecycles.

Since 1980 evolutionary biology has become an experimental science.

Adaptive radiation

The Caribbean islands host many versions of anole lizards. Usually, each island has at least three distinct types: a large arboreal lizard adapted for staying in the tree canopy, a ground lizard with long legs for running, and a bush variety that stays on low vegetation with short legs for navigation. One might conclude that an arboreal lizard species invaded Cuba, then Hispaniola, then Puerto Rico and others, and the adaptations for living in a canopy came from common ancestry. What actually happened is that an ancestral species of lizard landed on an island, and radiated into all the lizard niches on the island, and the same radiation was repeated on each island. The arboreal traits were convergently evolved many times on different islands.[167]

For years, the Caribbean anoles were the only such example, but recently that understanding has been changing. The snail *Mandarina* on the Japanese Ogasawara islands shows the same repeated radiation with some in the canopy, some semi-arboreal, and some on the ground. The adaptations happened repeatedly in parallel and fooled scientists until genetic studies confirmed that the snails are more closely related to their friends on the same island than their distant cousins that appear indistinguishable from them.[168]

A similar earth-shaking discovery came about recently with the bat genus *Myotis*. Rather than a species having evolved a single time and radiated to all similar species that dispersed around the world, DNA comparisons showed that differently-adapted bats living in the same regions were more closely related, indicating a repeated adaptive radiation.[169] Such examples have become commonplace since 2010.

Lizards on islands

Inspired by the guppy experiments, Jonathan Losos decided to run his own experiment on evolution.

He got the idea because in the mid-1970s another researcher left anole lizards on many lizard-free islands of various sizes in the Bahamas to test whether hurricanes would wipe out the population. In 1991, Losos returned and measured lizards from 14 of those tiny islands to see if there were any signs of the same adaptive radiation that was already known on the bigger islands. Did the anoles begin to diverge into arboreal, semi-arboreal, and ground lizards?

When the results of 161 anoles came in, the data showed as predicted. From the exact same ancestral population, within 14 years the lizards had diverged based on the habitat. There were short-legged anoles walking on thin branches, and longer legged anoles walking on the ground. Losos demonstrated that the adaptive radiation of lizards in the Caribbean not only showed repeated convergent adaptations, but that it happened in about as long as it takes to earn a PhD in evolutionary biology. Losos writes,

> After a decade of evolution, our fourteen populations differed in limb dimensions... Like the guppy studies, we could predict how lizards would evolve... re-create the conditions experienced by natural populations and the experimental populations would adapt repeatedly in the same way.[170]

Losos also addressed phenotypic plasticity. Were the changes a result of genetic variation? Or do the shorter legs come from growing up on thin branches? To quiet his critics, he raised two sets of the same lizards in different environments to see if the environment causes the legs to lengthen. To his surprise, the critics were right! There was a difference in leg length, indicating that some of the length change comes from usage, not genetic variation. However, as he describes,

> The lab growth experiment subjected the lizards to much greater differences – a narrow rod versus a broad piece of wood – than the differences in perch diameter among the experimental islands. Yet the difference in leg length among the island populations was three times greater than the difference produced in the lab… evolved genetic changes are likely responsible for the majority of the differences in limb length seen among the experimental island populations.[171]

Curly-tailed lizards on islands

One reason that some anoles have shorter legs is because some islands are inhabited by larger ground-dwelling curly-tailed lizards, and the anoles have an aversion to being eaten. Those with shorter legs are better adapted to the bushes and trees where the lunky lizard can't follow. But anoles enjoy the comfort of the ground and all the bugs that come along with it, so they can also grow longer legs to outrun predators and diverge into ground-anoles.

In 1997 Losos again visited the islands in the Bahamas and identified six islands for a control group, and another six islands each received five curly-tailed lizards. Three months later, the anole population was cut in half on the islands with the big lizards compared to that of the control islands. As predicted, after two years the anoles with predators were more often found elevated, preferring to perch much higher in the bushes or trees on narrow vegetation that favors short legs. But did their legs evolve to be shorter? Hurricane Floyd decided to end the experiment by inundating the islands and sweeping away every lizard.

However, when hurricane season was over, the scientists returned to find all the islands covered with baby anoles, hatched from buried eggs that survived the storm. The experiment was back on!

With all the anoles measured and new curly-tailed lizards introduced, they were ready to let nature happen again. Later, a hurricane removed all the lizards again, and again babies hatched from eggs and the experiment continued. They also ran some side experiments by placing arboreal anoles on small islands without trees and watching them adjust over four years until most of them, too, were swept away by a hurricane in 2012.

Eventually they were able to confirm their hypothesis: within just a few years, with the right selective pressure, anoles repeatedly showed the adaptive radiation into three species.[172]

One mouse, two mouse, red mouse, blue mouse

Evolution happens fast. Adaptations that were imagined over millennia are now known to take 2-10 years. The guppy and anole studies made the news and reverberated around the science community. Now fast evolution is the expectation, and just as the lizard experiment repurposed an older experiment, many others have gone back to re-analyze data created with other intentions to see if natural selection was in effect. Many are now running their own experiments. The ability to include DNA testing is a recent addition that has only enhanced the knowledge.

Dolph Schluter ran a stickleback (fish) experiment at the University of British Columbia with 13 artificial ponds of varying depths and showed several adaptive radiations in just two years

(Schluter also lost his entire experiment to the weather and had to start over once). Schluter helped in getting the stickleback genome partially sequenced for the first time in 2007, and he was able to accompany the physical changes with identified genetic variations.[173]

Color matching has long been associated with rapid evolution since the speckled moths of England adjusted to the industrial revolution. It has also been noticed all over the world as mice and other prey adjust to their background on lava flows that are only hundreds or maybe a few thousand years old. The reason is obvious and very well studied. Mice are eaten by many things, and a white mouse on a black background is much easier to consume, so the camouflaged mice live on to have the hanky panky that makes more of the same color.

It would seem rather pointless, then, to set up an enormous experiment in Nebraska with two huge steel enclosures 35 miles apart on different color ground. But that's exactly what Rowan Barrett did.

On 2.5 acres Barrett used 7,000 kg of sheet metal and chicken wire to create a grid of four enclosed spaces on dark soil, and he did it again on light soil. He confirmed that mice could not go under or over the walls, and then he dropped 100 mice in each enclosure. Half were from white sands and the other half from dark ground. It is not surprising that the white mice survived better on the white ground and the dark mice survived better on the dark ground. Their main predators were birds, after all. Considering the evolutionary experiments since the 1980s, it was also not surprising that the populations began to evolve and adapt in parallel, so that after five years the populations got either lighter or darker depending on the soil.

The reason this experiment stands out is that Barrett sequenced the genomes of all the mice before dropping them in the enclosures, focusing on a gene that is known to affect color variation in mice. At the end of the experiment, the populations were all sequenced again.

The team reviewed mutations that correlated with survival and began to articulate the genetic changes behind the rapid adaptations. They narrowed in on one particular amino acid that created a lighter coat when deleted. Lab testing with gene editing technology later confirmed that removing the amino acid caused visibly lighter color in mice. The preliminary results were published in a 1 February 2019 paper in *Science*, and the researchers say that they are continuing to analyze the data and will use the results to inform future experiments.[174]

With the addition of cheap genetic sequencing, evolutionary experiments will grow immensely over the next 20 years. They still appear to be in the very early stages of development and have immense potential.

Even faster evolution

The guppies and mice reproduce every 2-6 months, so their experiments require a long time to test natural selection. This, and the enormous resources required to set up and maintain the experiments make them difficult and expensive. They are also in nature where researchers have a harder time controlling variables (like, say, a hurricane).

E. coli bacteria, however, reproduce every 20 minutes and their environment is a warm test tube that can sit in a lab anywhere in the world. The genome of *E. coli* has been intensely studied for many years as a model organism. In 1988 Rich Lenski created 12 test tubes of genetically identical bacteria and subjected them to the same environmental stress. In 28 years, they went through 64,000 generations (it would take fruit flies 1,000 years to reproduce that many times).

Lenski published on his results frequently, amassing publications in prestigious periodicals and fame among evolutionary biologists. This is about as close as we've come to running Gould's experiment of restarting the clock on life to see if we'd end up with the same thing or something wildly different. The results tipped the scales toward convergence and repeatability. Lenski noted in 2011 at 50,000 generations,

> To my surprise, evolution was pretty repeatable...
> Although the lineages certainly diverged in many details, I was struck by the parallel trajectories of their evolution, with similar changes in so many phenotypic traits and even gene sequences that we examined.[175]

Similar to the guppy experiments inspiring many similar tests, the *E. coli* experiment on long-term evolution inspired perhaps hundreds of other microbial tests that are either newly published or ongoing as of 2022. Paul Rainey showed that the bacterium *P. fluorescens* performed the same repeated adaptive radiation into three species in just 10 days. Studies often find that the parallel adaptations are driven by exactly or nearly the same genetic changes, or in related genes with the same function. Yet there are still always genetic variations between parallel populations, and always a few exceptions.

Not so fast

The 12 *E. coli* populations had many identical genetic changes to the same environment, but there was one notable exception. One population developed the ability to digest citrate in the presence of oxygen, an ability that has never been documented in nature and only seen once in the lab before. This spawned its own side project that reanimated many older generations of that strain that were frozen for just such an occasion.

Five years and forty trillion cells later, researchers showed that the helpful adaptation was repeatable from older generations up to a point, but not from the original population of 1988. Was Gould right? Was the fate of evolution determined by chance and contingency? In the middle of this experiment, a revolution was going on, making DNA sequencing cheap and fast.

They went back and sequenced DNA at 29 points during the evolution of the citrate digesting strain, as well as the other strains. The ability to digest citrate in oxygen came from three particular mutations, and based on reruns from ancestral populations, the first two seem rare and the third one common. Thus, two seemingly rare mutations were required for the *E. coli* to make the needed change.[176]

The ability to reanimate and perform tests on the progenitor species is simply amazing. Every evolutionary biologist wishes they had a time machine to go back and test more primitive forms. Because of *E. coli*'s ability to return from frozen stasis, they can. The time travelling effect allowed the scientists to determine how likely a mutation was by turning back the clock and hitting play again, and again, and again.

Chance and inevitability and aliens

There seem to be a rush of experiments to test the value of chance and contingency vs convergence and determinism. The preponderance of the results demonstrate repeated, convergent, and fast adaptations. However, some beneficial adaptations take much longer and are less common (on a human timescale).

Many authors are trying to address the different philosophical viewpoints on chance and inevitability in evolution. The intellectual duals are battling for the heart of evolutionary theory. Stephen Gould is in the camp of contingency and chance, Simon Conway Morris is in the camp of determinism and inevitability, and others claim to be in the middle. They often quote the same examples with different conclusions. They usually start their works with a reference to Gould's thought experiment on replaying the tape of life, but there is another thought experiment that appears throughout the debate that is just as relevant: what will aliens look like?

One prominent example is in Edward Wilson's *The Meaning of Human Existence* (2015), where he tries to outline what features intelligent life would develop on other planets. According to Wilson,

- They would dwell on land so that fire could be used as an energy source.
- They would have a large brain in a head at the front of the body with sensory organs.
- They would rely on vision and sound for communication.
- They would have few limbs and at least one pair with pulpy tips for fine manipulation, and no claws.
- They would have small teeth and hunt with social cooperation and intelligence.
- They would have evolved in competing groups of the same species because groups dominated by selfishness need to be destroyed by groups dominated by altruism.
- They would have morality.

Wilson provides a useful outline of the qualities that appear necessary to develop civilization. He says that filmmakers portraying intelligent aliens with claws have got it wrong, but the aliens will most likely be humanoid. His views are in contrast to the more common view of contingency creating wildly different outcomes, such as Neil deGrasse Tyson, who thinks filmmakers get aliens wrong for the opposite reason:

[N]early all of them have two eyes, a nose, a mouth, two ears, a head, a neck, shoulders, arms, hands, fingers, a torso, two legs, two feet – and they can walk. From an anatomical view, these creatures are practically indistinguishable from humans, yet they are supposed to have come from another planet. If anything is certain, it is that life elsewhere in the universe, intelligent or otherwise, will look at least as exotic as some of Earth's own life forms.[177]

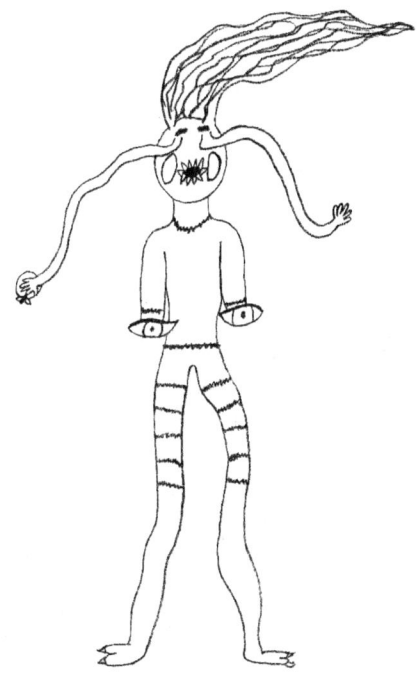

My 9-year old's attempt to draw an alien.

Could the two viewpoints be any more different? On one side, the strong push of evolution is broadly deterministic, evolution has repeatedly probed the likely outcomes, and something resembling humans is practically inevitable with the right environment. On the other, human intelligence has a meaninglessly small chance of forming, and "if anything is certain" intelligent aliens will not have any physical resemblance to humans. Jonathan Losos agreed that intelligent aliens would probably follow Wilson's description above, but he also said, "if any of a countless number of events had occurred differently in the past, *Homo sapiens* wouldn't have evolved. We were far from inevitable".[178]

They are looking at the same data. It's a philosophical debate about what are the likely outcomes of evolution and whether there is a human essence. We've come full circle, and now we're back to the views of 'Abdu'l-Baha.

The result of a text-to-image generator drawing an intelligent alien.

CHAPTER 7

THE HUMAN ESSENCE

Fundamental

On the subject of Darwinian theory, 'Abdu'l-Baha said, "Briefly, this question comes down to the originality or non-originality of the species, that is, whether the essence of the human species was fixed from the very origin or whether it subsequently came from the animals."[179]

This book so far attempts to determine whether or not 'Abdu'l-Baha believed that *Homo sapiens* has a common ancestor with living apes. However, 'Abdu'l-Baha's view of evolution appears focused on the spiritual implications of human origin, which can be embraced along with common descent. Likewise, a materialist could easily embrace a new model for the branching pattern of evolution and still regard human existence as a fluke of nature and without meaning. The real controversy is philosophical: Is existence overall random and accidental, or deterministic and fundamental?

This is the controversy that runs deep in the debate over science and religion. Those insisting that humans are an unlikely outcome are almost always outspoken atheists, to the point of ridiculing any scientist that believes in God. On the other hand, Simon Conway Morris, the outspoken champion of the deterministic view of evolution, is a Christian. From this debate over determinism immediately stems the question of the existence of God and the human soul.

An example of this debate can be seen in Robert Wright's 2009 article about what he considered "Brand Jesus". Wright is an agnostic who attempted to explain away religion. He said that Christianity became successful because it taught tolerance and amity across ethnic and national boundaries, "as a product of utility". He wrote,

> Why all the kin talk? Because Paul wasn't satisfied to just have a congregation in Corinth; he wanted to set up franchises – congregations of Jesus followers – in cities across the Roman Empire. These imperial aspirations, it turns out, infused Paul's preaching with an emphasis on brotherly love that it might never have acquired had Paul been content to run a single mom-and-pop store.[180]

In other words, in any given century on our planet, there must be thousands of people trying to start a religion. If the founder finds the right combination of principles that will draw adherents and form cohesive groups, that religion will be successful, and the others will fizzle out with time. Amity and kindness are essential qualities for any social movement to spread and be sustained, so any successful religion will have those qualities. He is describing religion in terms of biological evolution: the successful religions are conserved because they happened upon the right set of principles.

Wright describes the universality of human social principles as being the source of religion, as opposed to a divine source. Before any religious adherents get too carried away in protest, think again. He suggests that there are universal attributes of human virtue. Their preexistence may imply something about the universe.

Perfect

Consider the implications of 'justice'. Any creature that achieves a human level of consciousness will live in a society with language, and they will have a word for 'justice'. However, the mechanics of the language will be different, and the nature of laws and customs that try to implement justice will be different. Justice is an absolute concept, but the practical implementation of justice never achieves the absolute.

Another example could be drawn from the shape of a triangle or square-based pyramid. A perfect theoretical square-based pyramid exists as a concept, but its implementation cannot be perfect. The material used to compose it, the forces of gravity acting on it, and other superficial requirements of existence require that it never attains the ideal concept. The pyramid archetype is absolute; constructed pyramids can vary infinitely in details, but they can always be distinguished from any other polyhedron, such as a cube.

In nature, locusts appear in 13- or 17-year cycles, making it difficult for predators to evolve similar cycles. This is because those numbers are prime; only a multiple of itself and 1. The property of being prime is preexistent, and the fundamental attribute is found in biology. In the human field of mathematics, however, the property is entirely elaborated. In the movie *Contact* an alien race tries to contact earth with a radio signal repeating a sequence of prime numbers. The sequence of 2, 3, 5, 7, 11, 13, 17, 19, 23, 29 in radio bursts should be recognized by any intelligent life as being generated by another intelligent life form. This idea of communicating to alien races through mathematics was first proposed by Carl Sagan, and hints at the existence of universal concepts that would have to be used for communication with other planets.

Robert Wright described universal, preexistent principles that govern human relationships. The attributes of justice, love, kindness, cooperation, and unity are fundamental, absolute, universal, and innate. They exist above the natural world and its processes. They existed before they were expressed in a human mind. All humans have an attraction to them, whether consciously or not. The success of any endeavor is dependent on them. Of course, in the Baha'i Faith these are termed 'attributes of God', and these intangible realities are the aspirations of our social evolution. In the Baha'i teachings God is not a vertebrate with supernatural powers, but instead, all that can be known of God are these universal, eternal qualities, and periodically they are perfectly manifested in a supernatural way through a living human, representing the Founder of a religion.

Wright's understanding of the primacy of these principles was correct; social evolution drives mankind toward universal principles. Now we need to investigate if biological evolution drives species toward universal forms. Their preexistence may imply something about the universe.

Directed

The idea of suggesting a direction in evolution would make some scientists uncomfortable, because it is often associated with a supernatural direction. However, natural selection is pressure, which is a vector force that implies direction. In any given century on our planet, there must be uncountable genetic iterations. If the new code finds the right combination that will provide some advantage, the iteration of code will be successful, and all others will fizzle out with time. For example, the circulation of energy and the discharge of waste are essential qualities, so any successful complex organism will have them.

Convergence is commonly observed in nature. Biological creations approach a limit, just as constructed pyramids approach a limit of the perfect theoretical pyramid. In arguing against any goal in nature, a popular argument is that a monkey typing on a keyboard for an infinite amount of time could produce a work of Shakespeare; thus, enough genetic variation could produce any creature. The analogy has an obvious flaw by relying on the term 'infinite', and theists easily point to the fact that if there were as many monkeys as there are particles in the observable universe, and each typed 1,000 keystrokes per second for 100 times the life of the universe, the probability of a monkey producing a short book of Shakespeare would be effectively zero. Atheist author Richard Dawkins countered by pointing out that natural selection effectively prevents correct letters from changing, suggesting that a more accurate model would be a random phrase generator that keeps the letters that match the ideal target, then randomizes the remaining letters. In his test, the Hamlet phrase METHINKS IT IS LIKE A WEASEL was generated in 40 generations. I agree with Dawkins on this point. It is quite obvious from his example that nature, via natural selection, has an ideal target that is being selected for. A different lesson can be drawn from an actual attempt to run the monkey experiment. A keyboard given to six macaques at an English zoo produced pages consisting of mostly the letter S, followed by the keyboard being attacked and defecated upon. The director of the program said they had learned "an awful lot".[181]

This idea of direction in evolution toward preexistent forms is not such a new idea. The thrust of Keven Brown's *Evolution and Baha'i Belief* demonstrates that the concept goes back to the time of Plato. The Islamic philosophers continued debating the forms and essences of beings, and 'Abdu'l-Bahá's statements focus on emphasizing the absolute form to dispute the accidentalist direction that science was going in at the time.

Nor is the idea of an essence or spirit in organic beings such an unscientific one. Carl Woese, the noteworthy biologist, wrote in 2004 about science's failure to take a broad view of nature and determine what is accidental versus fundamental:

> Let's stop looking at the organism
> purely as a molecular machine...
>
> If they are not machines, then what are organisms? A metaphor far more to my liking is this. Imagine a child playing in a woodland stream, poking a stick into an eddy in the flowing current, thereby disrupting it. But the eddy quickly reforms. The child disperses it again. Again it reforms, and the fascinating game goes on. There you have it! Organisms are resilient patterns in a turbulent flow – patterns in an energy flow. A simple flow metaphor, of course, fails to capture much of what the organism is. None of our representations of organism capture it in its entirety. But the flow metaphor does begin to show us the organism's (and biology's) essence.[182]

Likewise, geneticist Eugene McCarthy wrote in 2008,

> Many physical, chemical, and mathematical systems tend toward particular states. When perturbed, a pendulum will eventually return to a stationary, plumb position. When the square root of any positive number is repeatedly taken, the series of results will converge on 1. When hydrogen and oxygen are combined, they

burn to form water. These stable states towards which processes tend are called absorptive. Thus, if a group of mating organisms is thought of as a process, what would constitute an absorptive state of that process?[183]

Essential

Recent authors have developed their own language to describe the repeatedly formed patterns in evolution. 'Ecomorph', 'form', 'evolutionary outcome', 'mutational pathways', 'archetype', 'recurrent pattern', 'derived functional states', 'ecological roles', 'configurations', these have all been used to describe the same basic concept, which 'Abdu'l-Baha and others have referred to as a 'species essence'.

For example, the thylacine and the wolf are commonly used as examples of convergent evolution because they both achieved the same form and filled the same ecological role, independent of inheritance. Try outlining the species essence of 'wolfness' and you'll see why biologists stuck with the most conservative and simple definition of a species.

Put descriptive boundaries on 'wolfness', and they will likely be either specific enough to be contradicted, or broad enough to be meaningless. The thylacine had the body structure and diet of a wolf, but appeared to be a solitary, ambush-style predator, whereas most dog-like species hunt in packs and chase their prey over some distance.[184] Most cat-like species use the ambush-style of hunting, but a few cats, like cheetahs or lions, chase their prey and hunt in packs. Using the definition of ambush-style for cats and pursuit-style for dogs is true generally, but breaks down in the particulars. Defining a species, then, uses a set of generalities that are almost always true.

Darwin acknowledges the confusion around defining a species, yet said, "every naturalist knows vaguely what he means when he speaks of a species".[185] The debate is reminiscent of US Supreme Court Justice Potter Stewart's response to the question of how to define pornography... "I know it when I see it."

It seems that defining a species essence is reminiscent of defining justice. Everyone knows vaguely what is meant by the term 'justice', but practical examples can be so ambiguous that society must rely on consensus. Even then, the examples are so flexible that laws need to be continually updated by a legislature and processed through the mind of a judge. No permanent descriptive limits can be defined, yet the principle of justice is eternal. Similarly, the essence of a species is real and permanent yet cannot be measured with instruments. The categorization of creatures into species essences also needs to be continually updated and processed through the mind of a scientist. This method of science-by-consensus may seem patently unscientific, but when Ernst Mayr traveled to New Guinea, he totaled up 136 species of local birds recognized by the natives. Western zoologists, using modern taxonomy, recognized 137. This convinced Mayr that a species is an objective fact. One of Mayr's pupils, Jerry Coyne, author and professor of Evolution at the University of Chicago, went on to write, "we know that species have an objective reality and are not simply arbitrary human constructs".[186]

Phenomenal

When a cactus grows from a seed, it relies on the affinity between atoms to organize molecular structures, which organize protein structures, which organize into organic material that can thrive and reproduce on earth. The affinities at every level are fundamental and repeatable. Life evolving on another planet with an earthly environment would eventually produce something similar to a cactus or succulent. The alien and earthly plants would share a form of kinship, because they are examples of the same perfect form or essence, continually being adjusted through natural selection to be more perfect in its particular essence. In a similar way, the pyramids of ancient Egypt, Mesoamerica and China share a form of kinship, because they were built as constructs of the pyramid archetype.

Now extend that principle of an essence to humans. Is there a fundamental biological organization that allows for intellectual powers? Clearly so, because these words are being read. It allowed primitive humans to break through a watershed when the species reached a certain level of biological development. That watershed is the development of imagination, thought, comprehension, memory, and scientific discovery, and they are strongly connected to the functioning of the brain. These attributes allow for the multi-generational education that builds a civilization. To a lesser extent these attributes can be seen in other animals, but in humans they are all perfected. These being described as beneficial traits that can be selected for by normal Darwinian evolution is really missing the point. There was a phase change, not more of the same. Consciousness is part of the available solution set. It is a result of natural processes.

The writings of 'Abdu'l-Baha describe the material composition drawing on an absolute biological principle that was preexistent, meaning it existed before the appearance of the species on our planet. 'Abdu'l-Baha said: "man has been man from his very inception and origin, and… the essence of his species has existed from eternity."[187]

The underlying principle of human essence is eternal and perfect, but its appearance on earth is phenomenal, and the phenomenal is changing and imperfect. A race of creatures on another planet may naturally evolve intelligence, but their chemistry and superficial details may be entirely different and adapted to a fundamentally different environment. Still, the capacity for natural processes to generate life and evolve intelligence is there, and any such life would convey the principle of justice, elucidate prime numbers, and build pyramids.

Preexistent

This idea of preexistence is perhaps the most abstruse idea related to the Baha'i view of evolution. Brown describes the species essence as time-invariant: "It is a natural law, universally valid for all times and all places."[188] The individual creature is not preexistent, it is created, phenomenal, and has some arbitrary elements. This is the view common to modern agnostic scientists, that our existence relies on contingencies of history. However, over the long arch of evolution, it is predictable and follows preexistent patterns. Preexistence implies something inherent to the universe, similar to the laws of physics and mathematics. 'Abdu'l-Baha, while describing preexistence vs. phenomena, says that,

> For the existence of each and every thing depends upon four causes: the efficient cause, the material cause, the formal cause, and the final cause. So this chair has a creator who is a carpenter, a matter which is wood, a form which is that of a chair, and a purpose which is to serve as a seat.[189]

He goes on to describe the spirit (also defined as 'essence') as preexistent and the body as phenomenal. The existence of the body depends upon the spirit, but the existence of the spirit does not depend upon the body. 'Abdu'l-Baha clearly taught that living creatures are phenomenal manifestations of a preexistent species essence, but how essences translate into specific examples has been debated for thousands of years, and will be debated for thousands more. An individual creature has unique characteristics and the species is always shifting and adapting, so how can the species essence be locked in place?

An analogy that may convey the variability of species is a pachinko board (aka bean machine), where beans cascading down an array of pins arrive at the bottom in a normal distribution. To make the analogy work, imagine a board in the shape of a cone with an acute angle to the ground. A bean is dropped on the apex and falls in a random direction down the face of the cone. As the bean falls to the first layer of pins, it takes one of two options, and each layer increasingly narrows the range of possible outcomes. The farther the bean falls, the wider the cone gets, and the more possible variations arise. As the bean makes its way down the pachinko board over millions of years, it arrives at the bottom with an outcome that gives the appearance of being one of an infinite range of possibilities. But the range of possible outcomes became increasingly narrow the longer the bean fell. After just a few rows of pins, the bean can no longer reach the pins on the opposite side of the cone. After a thousand rows, the bean has an incredibly narrow range of likely outcomes, but there is still variability within a range of outcomes.

In a similar way, a species gets its start by reaching multicellularity in one of a handful of developmental motifs, as described by Stuart Newman (see Ch. 4: Evolution of development). The motif broadly limits the possible outcomes and puts the species on a pachinko board for (perhaps) vertebrates, arthropods, cephalopods, or another animal phylum. At a very early stage in the evolution of a species, biological changes broadly limit possible results. A crab has gone down a biological path that will never evolve into a primate. Once these broad paths are set, the range of possible outcomes continues to become increasingly narrow, although still continuously changing.

'Abdu'l-Baha has said that species are original. That is *not* to say that the exact genome of a particular elephant was predestined to come into existence, nor does 'Abdu'l-Baha indicate what particular creatures are categorized in which species essence. But there is a normal distribution of outcomes within the species essence. The preexistence comes from the pachinko boards having been already set up when life began on the planet.

Meaningful

In the broadest terms, evolution is deterministic and will follow certain well-worn grooves. The process culminates in an intelligent being that achieves spiritual perception. The observable fact that the capacity for the human mind is inherent in the universe seems to give purpose to existence. Mindless processes of nature produced a fruit that is superior to nature, a creature that can deny instinct, manipulate nature to his own advantage, and express all the good virtues that exist only in social, not physical reality. It is this "distinction and capacity" to know and manifest these virtues that Baha'u'llah says, "must needs be regarded as the generating impulse and the primary purpose underlying the whole of creation".[190]

The social implications of such an attitude are very important. Richard Dawkins decided that the fundamental unit of selection is the loosely defined gene, as opposed to individuals or species, and he labelled genetic activity as "selfish". Even though he does not recommend drawing moral lessons from it, he describes humans as "survival machines – robot vehicles blindly programmed to preserve the selfish molecules known as genes". After reading *The Selfish Gene*, one Australian wrote, "at times I wish I could unread it... I largely blame *The Selfish Gene* for a series of bouts of depression I suffered from for more than a decade." Dawkins responded, "If something is true, no amount of wishful thinking can undo it." And that "there is indeed no purpose in the ultimate fate of the cosmos".[191]

The disagreement does not involve ignoring truth; there is a difference in the analysis of the same observable data. Genetic information exists because of its utility to progress towards a goal. The result is that certain genetic elements tend to be propagated because of their utility. That doesn't warrant them being described as 'selfish' elements, and our purpose being to carry them.

Dawkins' conclusion is antithetical to religion, not having anything to do with scientific analysis, but because his outlook on nature speaks to purpose in a way that is degrading to the human experience. If nature is described prominently as random, arbitrary, and selfish, then what does that say about humans, being the product of natural laws? On the other hand, if nature is correctly described as meaningful, perfect, preexistent, and cooperative, then humans will gain inspiration from reflecting on natural laws.

In a 2009 article, Simon Conway Morris, mentioned previously for his research into convergent evolution, wrote,

If, however, the universe is actually the product of a rational Mind and evolution is simply the search engine that in leading to sentience and consciousness allows us to discover the fundamental architecture of the universe – a point many mathematicians intuitively sense when they speak of the unreasonable effectiveness of mathematics – then things not only start to make much better sense, but they are also much more interesting. Farewell bleak nihilism; the cold assurances that all is meaningless. Of course, Darwin told us how to get there and by what mechanism, but neither why it is in the first place, nor how on earth we actually understand it.[192]

Created

A clear line should be drawn between the convergent, deterministic, stable, repeatable, multiple-origin theory being proposed and the ideas of intelligent design.

In 1925 John Scopes was convicted of violating a Tennessee law banning the teaching of evolution in classrooms. In the dramatic and public court trial that pitted science against religion, science just barely survived and evolution was allowed to be taught. Sixty-two short years later there was a full reversal of roles. In 1987 the US Supreme Court ruled that creation science could not be taught in public schools, citing the establishment clause of the constitution. Those wishing to promote a supernatural origin of life after the ruling changed their tagline from 'creationism' to 'intelligent design' and tried to introduce the idea into school curriculum.

In Dover, Pennsylvania the school board tried to require the teaching of intelligent design in classes. This culminated in a 2005 showdown in federal court, where proponents of intelligent design tried to explain evolution as an unverified theory. They claimed that biological construction was 'irreducibly complex', a term coined to indicate that biological order was so complicated that the only possible explanation is a special creation by an intelligent creator.

Earlier the same year another court battle was raging over intelligent design in Kansas. The Kansas school board decided to give equal class time to each of the theories of evolution and intelligent design, claiming that evolution was an unproven theory. In response, an Oregon State University graduate wrote an open letter to the school board, describing his own theory that the earth was created by a giant ball of spaghetti and meatballs, known as the Flying Spaghetti Monster. According to the author, this creator hid dinosaur fossils in the earth to fool people, and he wants people to dress and act like pirates, citing the correlation between declining piracy over the last two centuries and the rise in global warming. He wrote,

> I think we can all look forward to the time when these three theories are given equal time in our science classrooms across the country, and eventually the world; One third time for Intelligent Design, one third time for Flying Spaghetti Monsterism, and one third time for logical conjecture based on overwhelming observable evidence.[193]

It is easy to sympathize with the mockery of the blatant corruption of science, but some might also sympathize with the Discovery Institute, the primary driver behind the intelligent design movement, who said their goal was to "reverse the stifling dominance of the materialist worldview".[194]

Their approach has a subtle flaw; they want to find evidence that is consonant with their pre-determined views of science, they are not interested in unbiased truth. In this line of thinking, the intelligent design proponents latched on to any idea that vaguely supported their ideas or attacked their perceived enemies. In their desperation, the ideas were easily disproved, and they resorted to lying, threats of violence, and name-calling.[195]

I will be accused of the same flaw. Am I trying to support a pre-determined conclusion, or am I testing a hypothesis? Anyone challenging the scientific consensus to uphold the apparent meaning of 'Abdu'l-Baha's statements must grapple with this concept on a personal level, and must be ready to admit that the idea of parallel evolution is incorrect if confronted with overwhelming evidence.

Complex

Rather than focusing on universal and fundamental concepts in biology, the logic at the heart of intelligent design is that biological creatures are irreducibly complex, meaning that they couldn't have been formed by random variation involving chance. Therefore, proponents tend to fill any holes in science with the supernatural. A common example given during the Dover trial was the bacterial flagellum, which looks like a microscopic human motor, and is composed of approximately 40 specialized proteins. This structure, intelligent design proponents argued, is so precise that it must have been created with a plan already in mind. Other arguments by the Discovery Institute claimed that the Cambrian explosion could not have happened by natural processes.[196]

A similar logic can be introduced to the Great Pyramid of Giza. The complexity of its design is astounding, even by modern standards. The largest of the stones used weighed 75 tons, and were transported about 500 miles. The stones were cut so finely that a piece of paper can hardly fit between them. An understanding of the number of Pi may have been used during engineering. Even more confusing, when the pyramid was first broken into, it was apparently empty, with no mummy or treasure, and no evidence of torches having been used for light. There are certainly several unsolved mysteries about the Great Pyramid, and ancient alien theorists have been quick to fill in the holes with alien intervention.

It is best to make the fewest assumptions when using logic. If one hears the sound of hooves clicking on concrete outside a window, and it is known that horses frequent the streets, a simple explanation for the sound is that a horse is walking down the street. A conspiracy theorist may suggest that a person is standing outside the house clacking coconuts together, and then argue that they are equally valid theories, but they are not.

In the example of the pyramid, there is a long tradition of building larger and more complex tombs for the pharaohs of Egypt, the Great Pyramid being the largest and most complex. One theory is that the skills for construction evolved with time, as did the ego of the pharaohs who would be buried in them. The evidence or documentation of the methods of construction would have been destroyed with the passage of time and loss of cultural continuity. Another theory is that aliens came down and helped build the pyramids. Yes, both are theories, both are in the realm of what is possible, but one makes wild assumptions about an otherwise unobservable alien race, and the other makes virtually no assumptions. They are not equally valid theories.

Likewise, in looking at a bacterial flagellum without understanding how it evolved, one might make theories about its origin that involve all kinds of assumptions about special creation by a Flying Spaghetti Monster, but there is also a theory that supposes its gradual evolution from more simple forms following natural processes, a theory supported by observable phenomena and making very few assumptions.

In the debate over the flagellum, nobody bothered asking, 'How many times did it evolve?'

The Origin

Though I'm eschewing the creationist movement (as do many Christians), the Baha'i Faith does teach concepts associated with creationism, and they are pivotal to Baha'i belief. Baha'u'llah writes, "Every thing must needs have an origin and every building a builder."[197] He further describes God as the Fashioner, the Maker, the Creator, the Originator, the Origin,[198] and these titles are repeated and emphasized throughout His writings. 'Abdu'l-Baha said God "is preceded by no cause but rather is the Originator of the cause of causes".[199] And in another place regarding cause and effect, "Thus such a chain of causation goes on, and to maintain that this process goes on indefinitely is manifestly absurd. Thus such a chain of causation must of necessity lead eventually to Him who is… the Ultimate Cause."[200] In contrast with evangelical Christian creationists, the Baha'i teachings are clear that the physical universe has no beginning or end, and the story of creation found in Genesis is not to be taken literally. However, the Baha'i Faith clearly acknowledges the divine origin of the Bible, the source for those same Christian beliefs that now disagree with Baha'i beliefs.

Baha'u'llah addresses this issue, both directly and indirectly. He says that the stories found in scriptures regarding the beginning of creation are true, but that the 'beginning' is a reference to the revelation of divine knowledge. The nothingness that preceded the beginning is a reference to the ignorance that preceded human civilization. So, for example, where Baha'u'llah says that the energies endowed in man lie "latent within him, even as the flame is hidden within the candle", He further says that "until a fire is kindled the lamp will never be ignited".[201] The candle is lighted, He goes on to say, by the divine revelations of the Manifestations of God.

Humans of 200 thousand years ago have only minor physical differences compared with modern humans. They have the same capacity. The vivid contrast between the two is in the degree that their potential is manifested, via education. An ancient human raised in modern society could make breakthrough scientific discoveries, and likewise a modern human raised in the wild would revert to his animalistic nature and fail to manifest the virtues and intelligence latent within him. According to Baha'u'llah, it is the education brought by Manifestations of God that ignited the potential in the human candle. This is the creation described in Genesis. Baha'u'llah wrote,

As to thy question concerning the origin of creation. Know assuredly that God's creation hath existed from eternity, and will continue to exist forever. Its beginning hath had no beginning, and its end knoweth no end. His name, the Creator, presupposeth a creation, even as His title, the Lord of Men, must involve the existence of a servant.

As to those sayings, attributed to the Prophets of old, such as, "In the beginning was God; there was no creature to know Him," and "The Lord was alone; with no one to adore Him," the meaning of these and similar sayings is clear and evident, and should at no time be misapprehended. To this same truth bear witness these words which He hath revealed: "God was alone; there was none else besides Him. He will always remain what He hath ever been." Every discerning eye will readily perceive that the Lord is now manifest, yet there is none to recognize His glory...

Consider the hour at which the supreme Manifestation of God revealeth Himself unto men. Ere that hour cometh, the Ancient Being, Who is still unknown of men and hath not as yet given utterance to the Word of God, is Himself the All-Knower in a world devoid of any man that hath known Him. He is indeed the Creator without a creation. For at the very moment preceding His Revelation, each and every created thing shall be made to yield up its soul to God. This is indeed the Day of which it hath been written: "Whose shall be the Kingdom this Day?" And none can be found ready to answer![202]

This elaboration preserves the internal consistency of the Bible but also consistency between the Bible and the Baha'i Faith. Religious truth is not absolute, but relative to context, language, and culture. According to Shoghi Effendi, this relativity of truth is coupled with an understanding that "Divine Revelation is orderly, continuous and progressive".[203]

The stories of Genesis – and other creation myths – are interpreted as spiritual allegories, in a way that outwardly satisfied people's curiosity of history, while inwardly describing spiritual truths that uplifted society. The people of the time of Moses were preoccupied with spiritism and idol worship, with the spiritual and intellectual capacity of today's children. They generally believed that the world was an island continent on a flat disc with infinite water, surrounded by a dome with stars embedded in its surface. A modern understanding of how the earth developed was not possible or necessary at the time. The account of Genesis had the advantage of conveying essential spiritual truths with broad symbolism, and avoided the detail that was beyond their capacity to understand.

In a time when most people still clung to a literal interpretation of the story of Genesis, Baha'u'llah taught that the physical universe is without beginning or end. Yet like Moses, Baha'u'llah revealed new truths focusing on personal and social development, bringing about the highest level of civilization possible at the time.

CHAPTER 8

HUMAN FOSSILS

Recent human origin

New understandings of the genesis of life, of cellular evolution, and of the repeatability of biology provide strong support for the idea that some species do not share common ancestry. The descent of humans is another question and needs an investigation starting with modern humans and working backwards in time. Reconstructing history in this way relies primarily on fossil evidence, because even DNA analyses rely on assumptions that begin in the fossil record.

The subject of human origin was avoided by Darwin in his initial publication of *On the Origin of Species* in 1859, owing to the controversy he knew it would provoke. It wasn't until 1871 that he published *The Descent of Man*, making clear that his evolutionary theory should be applied to human origins.

Three years before Darwin's *On the Origin of Species*, a controversial fossil was discovered in the Neander Valley of Germany. It was a skull cap and other bones that looked human, but not exactly human, and they were fossilized, meaning they were very old

(50,000 years). This came to be termed Neanderthal and sparked a lot of scientific debate.

Neanderthal was understood to be a more primitive human, and when Darwin's theory also hit the scientific waves, there was a fevered race to find the missing link, defined as a half human, half ape fossil. The scientists of the time had an idea in mind of what the missing link would look like and expected a skull hosting a brain two-thirds the size of modern human, but with a chimpanzee jaw.

Soon the search began and a Dutch anthropologist by the name of Eugene Dubois set out for Indonesia, believing that wherever modern humans and apes coexist, there may be fossils of intermediary links. After years of failure, he stumbled upon a skullcap, femur, and teeth in 1891 that appeared to be from a very primitive human. They didn't quite fit the preconceived definition of the missing link, so Dubois changed the definition and declared that he had found the missing link. His claim was mostly dismissed back in Europe as an unknown ape variety. This Java Man (*Homo erectus*, 1.8 mya) turned out to be the first in a series of fossils found in Indonesia, China, India, Turkey, and eastern Africa that all appear to be part of a distinct species whose descent and progeny are still not entirely clear among anthropologists.

The next major discovery was in England. The scientific elites in England were having a culture war with Germany and France, and the discovery of Neanderthal in a German cave gave exclusive access to one of the greatest fossil finds to Germans. In 1912 'Abdu'l-Baha declared to an audience in San Francisco that "the lost link… will never be found".[204] Just two months later a British anthropologist presented Piltdown Man, a skull fragment and jawbone of a previously unknown early human. The brain size was two-thirds that of a modern human, and the jaw resembled that of a chimpanzee, with the exception of two human molars, showing the exact combination of features everyone expected to find in a missing link, a perfect transition between chimpanzee and human. Even better, the missing link was British. The response from the Royal Geological Society of London was overwhelmingly positive, and the case was closed on human evolution.

Thirteen years later, in 1925, Raymond Dart published his find in South Africa of an ancient human skull with a small brain and human jaw, exactly the opposite of Piltdown Man. The new find, called Taung Child (*Australopithecus africanus*, 2.8 mya), was dismissed by British experts as a gorilla because it conflicted with their view of human evolution embodied by Piltdown Man. Dart was an Australian, not part of the establishment; he found the fossil in South Africa, not Europe; and the find had a small brain, a feature inconsistent with the establishment's views. It wasn't until 1953 that a thorough investigation of the Piltdown fossils was undertaken, and it didn't take long to find that the bones were not fossilized, the skull was human, the jaw was of an orangutan, and the filed-down teeth were from a chimpanzee, all of which were stained to look old. It was the most elaborate hoax in anthropological history, and it had led scientists down a blind alley for 41 years.

By the late 1950s, new hypotheses appeared among researchers about the causal relationships that led to modern humans. The new narrative was that tool use was the key to the human jaw, which led to the brain development that distinguishes *Homo sapiens*. In 1959 a new fossil, Zinj (*Paranthropus boisei*, 1.75 mya) was found in Kenya by Mary and Louis Leakey among a layer of tools. But Zinj had incredibly powerful jaws, indicating that it didn't use tools. Louis announced that Zinj was the missing link. The following year Mary and Louis found another fossil in the same place, but an entirely different species without the powerful jaw (*Homo habilis*, 1.75 mya), so the new species was announced as the missing link, and Zinj was relegated to an extinct species. This created another paradigm shift in human evolutionary thought. Two human-like but distinct species of early hominins both co-existed in the same place at the same time.

What had been a single line of descent was replaced by a series of lines. Between 1925 and 1965, over a hundred fossils were found in South Africa alone. All can be dated with increasing accuracy to form a line of descent of several species, some being evolutionary dead ends, others being transitional toward a modern form. A tree began to form of the genus *Homo*, and genetic analysis has shown extensive hybridization among them.

In the early 1970s a team of researchers began searching in Ethiopia for an older common ancestor to the Taung Child (2 mya) and *Homo habilis* (1.75 mya), which were on separate ancestral lines. They chose a rock layer of about 3.5 mya and began searching. Soon they discovered 40% of a fossilized skeleton, the most complete skeletal remains of the oldest human ancestor yet discovered, dated to 3.2 mya. She was dubbed Lucy, as "Lucy in the Sky with Diamonds" blasted on the radio in the camp. Lucy (*Australopithecus afarensis*) filled a hole in the human puzzle by explaining how humans came to walk on two feet. The savannah hypothesis stated that as trees became more sparse, human ancestors needed to move on the ground between trees. Standing upright gave an advantage to seeing predators coming through the grass. Walking upright freed up the hands, which allowed for tools, which allowed for brain growth, which distinguished modern humans. The case was closed on human evolution.

Move over, Lucy

The savannah theory was befuddled in 1999 with the publication of a collection of articles in the journal *Science* about Ardi (*Ardipithecus ramus*), initially discovered in 1994, the new oldest fossilized hominin ever discovered at about 4.4 mya in Ethiopia.

It wasn't until 2009 that a comprehensive analysis of the implications of the genus *Ardipithecus* was published in the journal *Science*. The discovery of Ardi pushed back the date of bipedalism, and with it the estimated common ancestor with chimpanzees to at least 6 mya, 2 million years earlier than previously thought.[205] The discovery also concluded that the savannah hypothesis did not explain bipedalism, because Ardi walked upright while living in dense forest.

Ardi had some primate features, but also shared some features exclusive to modern humans. Also surprisingly, Ardi does not share certain features common to modern chimpanzees, such as knuckle walking, leading researchers to believe that the common ancestor of modern chimpanzees and humans did not resemble either.[206]

Ardi has put to rest the notion of a missing link as theorized since the time of Darwin, resembling something between humans and today's apes. As *National Geographic* put it, "Indeed, the new evidence suggests that the study of chimpanzee anatomy and behavior – long used to infer the nature of the earliest human ancestors – is largely irrelevant to understanding our beginnings."[207]

For example, it was long believed that knuckle walking was a trait held by the common ancestor of all living apes, because humans have evidence in wrist and hand bones shared by African apes that would support the theory, but Ardi did not knuckle walk. A detailed 2009 examination of wrist bones of several primates caused evolutionary anthropologists to re-evaluate what they'd said for years. Now knuckle walking is believed to be convergent evolution, developed independently in chimps and gorillas, and the human remnants of fused wrist bones are no longer considered trace members because lemurs have the same feature and do not knuckle walk.[208] From 1912 until the full publication in 2009 of Ardi's discovery, the overwhelming consensus among scientists conflicted with 'Abdu'l-Bahá's statement that the missing link would never be found. For the first 41 of those years, the primary evidence turned out to be an elaborate hoax.

What of the chimpanzee fossils? You may be surprised to know that there are none. That's right, none. With the exception of some teeth found in Kenya in 2005 dating to 545–284 thousand years ago, no fossilized remains have been found of a proto-chimpanzee, and only fragments of proto-gorillas.[209] Humid forests don't fossilize bones very well, because predators or bacteria would eat any dead animal before it had a chance to become a fossil. The result is a bias in fossil analysis that favors certain kinds of organisms over others,[210] and fossil evidence is the only direct evidence of extinct species and their evolution.[211] Consider that observers once posited that ancient mariners never left the shoreline of the Mediterranean, because the observers only found ship remains along the shoreline, which happened to be the only place they could observe.[212] Later, underwater archaeology dissolved that argument when shipwrecks were found in deeper waters. Observers of early human fossils may be treading similar water if they suggest that humans only lived where fossils have been found. Articles commonly point out that the oldest hominin fossils begin to appear at the same time and place as the savannahs appeared in eastern Africa; they then cite this correlation as evidence that the human line appeared because of the savannahs.[213]

The recent discovery of chimpanzee teeth was another nail in the coffin of the savannah hypothesis to explain bipedalism, because they were coexistent with a species of *Homo*, and they were found in the eastern Rift Valley, meaning that the valley was not an impenetrable barrier to chimpanzees, as previously believed.[214]

With the logic behind the savannah hypothesis discredited, there now exists no comprehensive theory that explains a human trait that is unique among all earthly primates. One idea suggests that humans developed bipedalism while walking on tree limbs for stability, similar to behavior observed in orangutans. Another idea suggests walking bipedally conserves energy. Another suggests that humans had a recent amphibious origin, because the only time apes definitively walk bipedally is in shallow water.

Move over, Ardi

Although Ardi remains the oldest semi-complete skeleton of a hominin, 20 specimens of a 6-million-year-old bipedal species were found in 2000 that came to be known as the genus *Orrorin*. This was announced by its discoverers to be the earliest fossil of a human ancestor.

Then in 2002 a new discovery complicated matters. A skull was found near Lake Chad, whose analysis came out in a 2006 article of *Nature*. Named Toumaï (*Sahelanthropus tchadensis*), the skull had features indicating a likely bipedal, potential ancestor of humans, but dated to between 7.2 and 6.8 mya.[215] It is still unclear to researchers whether the Toumaï skull was part of the two species after the split, or whether it is a descendant of neither: an extinct species with no modern descendants.

The problem with Toumaï is that DNA analysis has placed the separation of chimps and humans at 6.3-5.4 mya,[216] the divergence of gorillas at 8 mya, and the divergence of orangutans at 12-16 mya.[217] Toumaï has "changed substantially the understanding of early human evolution in Africa", establishes "new calibrations of the molecular clock", and "testifies that the last divergence between chimps and humans is certainly not much more recent than 8 [mya]".[218] As the date of speciation moves back, it creates further confusion by bumping up against the estimated gorilla divergence, previously assumed to be 8 mya, but recently revised to be 10 mya.[219]

Hominin descent

A 2006 review of the Ardi findings noted that, "the strongest calibration is now from hominins themselves... whose derived characters effectively falsify late divergence estimates".[220] The implication is that fossil evidence trumps molecular DNA estimates, which are so flexible that a divergence estimate can be adjusted by 100% without much soul searching.

In the face of fossil evidence, scientists changed their definition of the common ancestor. A 2006 article in *Nature* took another look at the DNA evidence, and decided that the time from the beginning to the completion of the separation of the species ranged over 4 million years, requiring an initial split followed by several hybridization events.[221] This dilemma has remained unresolved, and its magnitude mostly ignored by the mainstream. One of the first articles published about Toumaï simply invoked the principle of evolutionary convergence to explain away any findings that contradict the standard narrative.[222] Clearly some researchers pick and choose when describing similar traits as either convergent or derived, simply reinforcing what is already assumed.

Move over, Toumaï

3.7 million years ago a volcano erupted near Laetoli, Tanzania and left a thick layer of ash on a field. When things settled down, two human-ish adults took a leisurely stroll through the field, leaving impressions of their feet in the ash. Sometime after that, a smaller individual walked through the same footprints. Then some light rain fell on the field, hardening the prints into cement. Then another layer of ash fell on the dry cement, preserving the prints until Mary Leakey uncovered them in 1976. To everyone's amazement, the gait was almost indistinguishable from that of a modern human, fully upright and driven by the front of the foot. The same layer of ash also held fossils of *Australopithecus*. This was the oldest known example of human-like footprints until 2017.

In 2002 Gerard Gierlinski took a break from paleontology and vacationed in Trachilos, Crete, where he took note of some fossilized mammal tracks. He eventually returned in 2010 to investigate, and to his surprise, they were unmistakably human-like and dated reliably to 5.7 mya. Crete is now an island in the Mediterranean, but when the prints were made it was connected to Greece, and a vast forest covered the Sahara and eastern Mediterranean shore.[223]

The Cretan trackmaker lived before *Ardipithecus* (Ardi) and after *Sahelanthropus* (Toumaï), which is surprising because Ardi's toe pointed to the side, like a chimpanzee, indicating that *Ardipithecus* may not be a progenitor genus of humans. The find was also surprising because the oldest evidence of hominins outside of Africa was previously 1.9 mya.

When Gierlinski published his findings in 2017, he noted that, "The interpretation of these footprints is potentially controversial" and that the trackmaker was either an early member of the human line or a "primate that convergently evolved human-like foot anatomy".[224]

Coincidentally, in 2017 another discovery was being published about early humans in Europe. In 1944 Nazi forces were building a bunker in Greece and found a fossil jawbone that was set aside for later study. The jaw was recently associated with a premolar from Bulgaria dating to the same period, both of which have features characteristic of modern humans and pre-humans such as *Ardipi-*

thecus and *Australopithecus*, and date to 7.2 mya. The researchers named the genus *Graecopithecus* and concluded that it represents the first hominin after the split from the common chimpanzee ancestor.[225] Another piece of the European origin story came from a pelvis found in 2006 in Hungary indicating an upright-walking, tree-dwelling, textbook-rewriting ape of the genus *Rudapithecus* that dates to 10 mya.[226]

Beyond these examples, there are many other fragments of ape fossils and much speculation about relationships. The full picture is messy, especially trying to describe a human ancestor before upright walking and forward facing toes. For example, the femur of bipedal *Orrorin* is most similar to a 20-million-year-old ape *Ekembo* from Kenya, suggesting that the most recent common ancestor of apes and humans might be among the Miocene apes, between 23 and 5 million years ago.[227] The oldest primate fossil is a tiny creature, smaller than your fist, who lived in Wyoming 56 million years ago.[228]

Special species

Discoveries continually support the idea that human distinction goes back farther than previously imagined. Importantly, they lack a clear line of fossils connecting humans and chimpanzees, leaving possible the confirmation of 'Abdu'l-Bahá's statements.

I believe further discoveries will only further confound the traditional view. The trend is that new discoveries cause dramatic shifts in the understanding of human evolution. The next few decades will see more major shifts as humans take a peek into their origins, especially because the technology for fossil discovery and analysis improved by leaps and bounds since the 1990s as the computer age dawned. Perhaps there will be a discovery of ancient humans in Asia; perhaps another previously unknown line of human lineage from a surprising time in a surprising place; perhaps soft tissue fossils showing something unexpected.

Further discoveries may show the human line evolving from a more primitive form, and at a certain point that form may be outwardly undistinguishable from other animals. At that stage, relationships will be extremely difficult to follow, as demonstrated by any fossil finds before bipedalism was adopted. Upright walking seems to be the oldest uniquely human morphological trait.

The current fossil evidence supports the hypotheses that species evolve and adapt by natural selection within evolutionary constraints, producing repeatable convergent traits among similar forms in parallel, and evolving from a primitive form. The evidence for the human line sharing a common ancestor with a non-human form is more vague. Perhaps if the chimpanzee lineage were present in great detail in the fossil record, the evidence would be more clear, but I cannot conclude that fossil evidence disputes my understanding of 'Abdu'l-Baha's comments. Scientific discovery has only moved in the direction of agreement. My hypothesis, however, still conflicts with genetic evidence of common ancestry.

A tale of two orangutans

The chimpanzee has long been regarded in mainstream science as the closest living relative to humans, and very few authoritative sources have proposed any other option. One such attempt came in 2005 by Jeffrey Schwartz, a professor of anthropology at the University of Pittsburgh, with his publication of *The Red Ape*. He was also joined in 2009 by John Grehan of the Buffalo Museum of Science to co-author an article in the *Journal of Biogeography*. They conducted research that rejects the popular suggestion of chimpanzees being the closest relative to humans, instead suggesting orangutans.

Peter Andrews, paleoanthropologist and author on human evolution, reviewed the work and said that the authors provided good evidence to support the theory, "It must be taken seriously, and if it reopens the debate between molecular biologists and morphologists, so much the better."[229]

Andrews is hinting at the two methods of establishing relationships: physical and fossil evidence (morphology) and DNA similarity (molecular biology). With the tidal wave of data coming from DNA sequencing, the well-worn path of morphology has been all but forgotten. Schwartz and Grehan specialize in the side of physical anthropology. According to them, humans are physically most similar to orangutans, and fossil evidence supports a human-orangutan connection.

Their works cite 63 physical characteristics that are unique to great apes, 28 of which are shared between humans and orangutans, compared to two features shared with chimpanzees, and seven with gorillas. Significant among these was the thick enamel found on the teeth of only humans and orangutans, but no other apes. In addition, early human fossils, such as Lucy, become increasingly similar to orangutans as they get more ancient.

The authors proposed that humans and orangutans shared a common ancestor at least 12 mya. The common ancestor would have lived during a period of thick forest extending from Asia to Africa that was later broken up by changing ecosystems, which left the human line in Africa and the orangutan line in Southeast Asia. The authors point out that a presumed human-chimp relationship lacks fossil evidence and requires a series of "complicated and convoluted" scenarios where early apes began in Africa, migrated to Europe leaving none in Africa, then diverged with some moving to Asia, and others back to Africa, leaving Europe without apes.[230]

They also provide a detailed criticism of assumptions behind DNA comparisons, concluding that the molecular studies showing a human-chimp connection are flawed. They point to four problems: ignoring genetic convergence, insufficient sampling, exclusion of the orangutan, and inconsistency with physical evidence. They further conclude that when presented with inconsistencies between morphology and molecular data, that physical evidence should win in a fight between the two.[231] In the case of human origin, physical evidence is not pointing to the chimpanzee.

Sequencing DNA became relatively inexpensive by 2010. That year a study of several orangutan genomes presented the "surprising finding" that some regions of the human genome are more like orangutans than chimpanzees.[232] A separate review in *Science* noted that humans and orangutans share structural DNA similarities that chimps and humans do not, leading the author to conclude that chimps lost the attributes after their divergence.[233]

The fully sequenced gorilla genome, available in 2012, led researchers to conclude that there are significant genetic similarities achieved convergently in different ape species. For example, the genetic changes to develop hearing are very similar between humans and gorillas, but different in chimpanzees. In many areas of the genome chimpanzees appear to be more closely related to gorillas than humans. Altogether, "15% of the human genome is closer to the gorilla genome than it is to chimpanzee, and 15% of the chimpanzee genome is closer to the gorilla than human".[234]

The most striking finding was the presence of similar chromosome caps on gorillas, chimpanzees, and other apes such as gibbons, but they were absent from humans and orangutans, leaving the authors to conclude that the similar caps evolved independently and convergently in several species of apes.

Why all this talk of orangutans while discussing the independent descent of humans? Showing orangutan-human common ancestry refutes the idea as much as chimp-human common ancestry. There are two important concepts that are relevant.

First, this debate demonstrates just how difficult it is to pin down the descent of any species. It is not a matter of simple observation; it requires logic and assumptions. Morphology should be regarded as the primary 'hard' evidence, since even DNA analyses rely on assumptions that begin in the fossil record.

Second, the authors of the orangutan study provide authoritative and logical criticism of blatant assumptions in DNA comparison. These kinds of critical analyses of the orthodox position on molecular data are hard to come by outside of creationist authors.

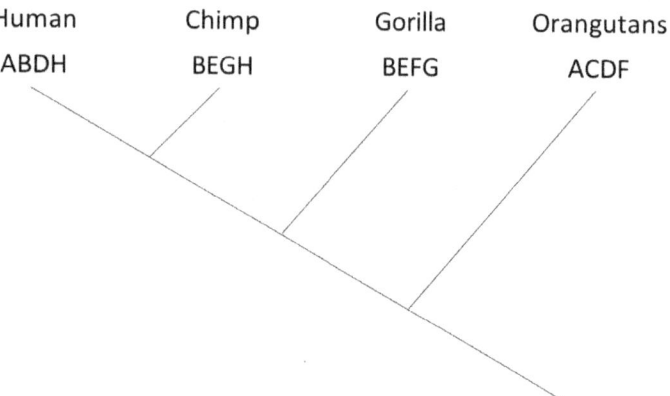

Human ABDH   Chimp BEGH   Gorilla BEFG   Orangutans ACDF

Study the sequences above and try to identify how species diverged. The consensus scenario is presented for ancestry. This oversimplified example with arbitrary letters reflects recent discoveries among the apes. Humans and orangutans share features that are absent in chimps and gorillas (AD), and chimps and gorillas share features that are absent in humans and orangutans (EG). The branching pattern requires convergence or horizontal gene transfer to explain inconsistency. Was 'A' added to both the orangutan and human lines? Or was it deleted once on the gorilla line and again on the chimp line?

Both hypotheses argue for common descent with either chimpanzees or orangutans, but underlying both is an unquestioned belief that divergence from a common ancestor is the only source of speciation. Of course, there is another option to explain the data at hand. Independent descent with a high level of constraint and convergence is the best explanation for the origination of disparate phyla, but it might also apply all the way down to the genus or family. Maybe when your tool is a hammer, every problem is a nail, and when your only tool to explain speciation is divergence from a recent common ancestor, every species needs to be placed on the tree of life regardless of the data at hand.

The DNA evidence, however, remains the clearest proof of common ancestry, and there are three specific topics that need to be addressed to support the hypothesis that humans and chimps may not share a recent common ancestor.

- General overall DNA similarity, including non-functional sections.
- Chromosome fusion that formed human chromosome number two.
- Endogenous retroviruses that appear in the same locations in humans and chimpanzees, being described by many as the most compelling evidence for common ancestry.

Any theory claiming to admonish the established doctrine must address these.

CHAPTER 9

GENOMES

With a few exceptions, I consolidated all the information on DNA comparisons into this chapter. It is a great challenge to write for a wide audience, from my mother who isn't sure what a genome is, to someone with a degree in genetics. If you're very familiar with genetics, skip over the DNA basics section. If you're like my mother, feel free to skip over this chapter and jump to chapter 10, you won't miss a beat.

DNA basics

The discovery of the double helix of deoxyribonucleic acid (DNA) forever changed humanity's view of biology. Where previously there was a mysterious process, science elucidated a molecular machine. Reading the code opened a new approach to evolution as genomes could be compared.

DNA is made up of four nucleic acids (A, T, C, and G), also called base pairs or nucleotides. Base pairs act like letters, three letters form codons that act like words, a series of words form genes that act like sentences, each sentence is surrounded by regulatory sequences that act like paragraphs, each paragraph is grouped into chromosomes that act like chapters, and each chapter is grouped with many others inside a nucleus to make the book of the genome.

A human nucleus has over three billion base pairs, the equivalent of about 770MB of hard drive space, and among the data there are about 25 thousand genes.

One of the 64 possible codons identifies the beginning of a gene, and three of them can identify the end of a gene. In between, each codon represents a single amino acid, with 20 amino acids that can be coded. The string of amino acids goes through a compiler that fits them together to form a protein, and the protein goes on to perform some function, for example, to move salt between a membrane.

Within genes, changing a single letter can change the amino acid, which can change the protein and the function. If the function is vital, for example, if the protein processes energy in the brain, then any babies born with the mutation will die and the new iteration of the gene would die with it. If, however, the mutation turns out to be helpful, like providing more efficient energy processing in the brain, then its owner would be more likely to pass it on, and the gene would be propagated. In between are many mutations that have no fatal effect, but may affect the organism for better or worse later in life or during hardship, such as a famine.

Proteins are three-dimensional. That string of amino acids must be folded, and the shape changes the function dramatically. Some genes are turned on and off throughout life, or in different environments, or in different cells, and some traits are recessive. Outside of the areas identified as genes, there are long stretches of DNA whose function is not clear. Then only two decades ago a new set of molecular interactions was discovered affecting proteins and heritable traits that stand outside of the DNA, now referred to as epigenetics.

Enormous strings of DNA are packaged into chromosomes (humans have 46) that take a three-dimensional shape, the structure and function of which is quite complex. Mating populations must share the same general number and structure of chromosomes (karyotype), so that the genetic information can be used in reproduction to create a viable offspring. A cell carries DNA in both the mitochondria and the nucleus, so references to DNA should clarify which variety is being discussed.

The variations at the heart of natural selection can be broken down into the following categories.

- Point mutations can change, add, or delete individual letters while cells make copies of themselves; these represent perhaps 60 mutations per individual.[235]
- Chromosomal mutations can change the number and structure of chromosomes; these represent perhaps one per thousand births.
- Retroviruses add their own DNA into other cells.
- Sexual reproduction produces a mixture of the parents' DNA.
- Hybridization between disparate species produces new karyotypes.
- Transposable elements are cellular functions that sometimes go around copy and pasting, or cut and pasting sections of code.

For simplicity, all these can be referred to as mutations, although the most dramatic mutations are structural changes to the chromosomes themselves, often through hybridization but also from copy errors or transposition events. Furthermore, the various cells in a body mutate independently of each other, so an organism is a patchwork of minor DNA iterations. That means that the top of a tree has different DNA from the bottom of the tree.

If mutations produce a negative result, lessening its selective advantage, then the mutations will not be passed along over evolutionary time, and the original sequence is said to be 'conserved' or 'constrained', because it becomes resistant to change.

Junk

In 2001 the human genome was first published in draft form, and some of the first comments were that only 1.5% of human DNA coded for proteins (genes), while the rest was 'junk DNA' left over from evolution with no discernible function and no selective value. A rushed comparison to the other genomes showed similarities in the non-coding DNA, and these were presented as irrefutable proof of recent common ancestry.

'Abdu'l-Baha already answered this argument. In his day it came in the form of common physical traits that were thought to be useless. The scientists of a century ago proposed trace members of organs as proof that humans share a common ancestor with primates. The human coccyx (tailbone) is a classic example of a vestige – a useless remnant of once owning a tail. The same remnant appears in all the tailless apes. 'Abdu'l-Baha agreed that humans evolved from a primitive form, but he clearly argued against vestiges as evidence for common ancestry. He said,

> ... these minor traces and vestigial limbs might have some great underlying wisdom which the human mind has so far been unable to fathom... the blackness of the pupil of the eye is due to its absorbing the rays of the sun, for if it were another colour – say, uniformly white – it would not absorb these rays. Now, so long as the wisdom underlying the things that we have mentioned is unknown, one may well imagine that the reason and wisdom of the vestigial limbs, whether in the animal or in man, is also unknown.[236]

This should sound familiar. It is the same logic used by modern scientists promoting determinism and convergence in evolution.

The blackness of the pupil is one of many attributes required for a functional camera eye and it has appeared independently numerous times in evolution. The repeated appearance of the organ implies determinism, and when there is only one way to do it, the similarities do not necessarily imply common ancestry. As soon as function and purpose can be attributed to something, convergent evolution can generally explain similarities in different species without relying on derivation from a recent common ancestor.

'Abdu'l-Baha suggested that trace organs might not actually be useless. The coccyx in its current form does, in fact, provide a useful function. Just ask anyone who breaks it or has it surgically removed due to problems. The coccyx provides support and stability while sitting, acting as the leg of a tripod. It also serves as the attachment site for numerous muscles and ligaments. Similar arguments can be made for other proposed vestigial organs.

Now applying this to junk DNA, it is precisely the presumed non-function of the junk DNA that makes it strong evidence for common descent. Mutations happen randomly and undirected mutations that don't go through a selection process would not converge on near-identical sections of code. But when function is ascribed, the evidence for ancestry becomes more vague, because an attribute that provides value will be resistant to change and can be selected for independently in unrelated organisms. Is most of the genome functional or vestigial?

ENCODE

There is strong selective pressure to reduce unnecessary genetic material, because it takes time and energy to maintain DNA, especially when making copies of cells. In fact, it has long been known that organisms with the longest genomes are more likely to go extinct.[237] This principle should assume function to DNA unless nonfunction can be confirmed, which is incredibly difficult.

Regarding the sections of apparently non-functional DNA, starting in 2005 and accelerating in 2009 the junk DNA was found to be not just functional, but invaluable. At least 5% of the mammalian DNA was found to be necessary for life,[238] and the majority of it is highly conserved or picked up by messengers, implying an unidentified function.[239] Some are switches that turn genes on and off, some tell when or where genes are expressed, other sections support the physical structure of the chromosome. Various theories have been proposed about the evolutionary role of repeated sections of code, such as stability, efficiency, or the ability to evolve;[240] including the structural view, the majority of DNA is best described as 'unknown function', rather than 'no function'.

With DNA being recorded and observed in larger quantities, a consortium formed that began the arduous task of documenting what the human genome actually does, as opposed to just reading off the letters. This Encyclopedia of DNA Elements (ENCODE) brought together over 400 scientists in 32 countries to take on the task. During the summer of 2012 they published a series of papers documenting the results, which were perhaps more revolutionary than the original human genome project. Of course, they found that the human genome is mostly functional and alive, akin to a jungle with complex interacting parts. There were several noteworthy conclusions: 75% of the genome is transcribed into RNA at some stage, implying that it gets used somehow, 62% form transcripts that look stable, and four million switches control 20,000 genes in 147 cell types.[241] A further study discovered a new type of RNA and confirmed in 2013 that about 85% of non-coding human DNA is transcribed into RNA. It is still a bit premature, and difficult to define 'functional', but it appears that the majority of the human genome is functional, and there remains no end in sight for further research into how things work.[242]

The label of 'junk DNA' for 98% of the human genome will go down in history as one of several cases of incorrect presumption that was later shown false.

Lactose tolerance

While living in China, I tasted many unfamiliar foods. Jellyfish tentacles, dog meat, delicious frog legs, pigeon, and eel, just to name a few. Of course I also tried the ice cream, and it left my mouth unusually oily. So I investigated. Among all the strange-to-me things Asians eat, animal milk is not one of them, so they don't digest lactose well as adults. Instead of milk cream, their ice cream is based on almond or coconut cream.

A normal human can properly digest lactose until about the age of five. After that, the gene that makes the digesting enzyme (lactase) turns off, because what human would continue nursing after the age of five? As it turns out, many humans nurse through-out their whole lives, from a cow. The human populations where this practice was widespread over thousands of years developed both lactose tolerance (lactase persistence) and very tasty desserts. Instead of their gene turning off around the age of five, theirs was permanently switched 'on', producing the lactose-digesting enzyme their whole lives.

There are precious few examples of natural selection among modern humans. Supporters of evolution have long pointed to lactose tolerance as evidence against Intelligent Design. They were right, but they should have gone further and continued probing. How many times did natural selection produce this result?

When the human genome was published, lactose tolerance was the target of several genetic investigations. What researchers found was that the genetic change behind lactose tolerance was largely a single nucleotide, not in the gene that produces the lactose-digesting enzyme, but in 14,000 base pairs upstream from it in a region that controls activation of the gene. This finding emphasizes the selective value of regulatory regions previously called junk. Further, the actual nucleotide change varies among populations, with four variations of the mutation that evolved independently, as confirmed in a 2007 study.[243] It is a marked example of convergent evolution. But even this study should have gone further and continued probing.

A second study in September of the same year found that one of those four mutations (T-13910) actually emerged at least twice. Not only has the exact same nucleotide change occurred more than once in cow-loving populations, but the convergent adaptation is an ongoing process.[244]

One of the few examples of recent human evolution shows that lactose tolerance has been achieved convergently several times among diverse populations; the genomic changes behind the adaptation largely hit the exact same region within a few nucleotides; one particular nucleotide change has been made independently several times; and the true level of genetic convergence is entirely masked when the genetic changes are exactly the same.

Genetic convergence

Like phenotypic convergence, only trivial and isolated examples of genetic change have traditionally been declared as convergent adaptation, and only when no other explanation exists. Like phenotypic convergence, a steady trickle of examples of genetic convergence is gradually shifting the paradigm, and many researchers are beginning to admit that on a much larger scale than previously thought, genetic similarity can be attributed to high constraint and molecular convergence.

An early dramatic change came from Lenski's *E. Coli* populations (see Ch. 6: Even faster evolution) that started in 1988 and produced many publications over many years about the similar parallel growth of 12 populations. Lenski inspired many other experiments.

Researchers as far back as 1994 wrote, "Given the difficulties associated with alignment and with establishing the conditions of consistency and convergence, it is clear that molecular phylogenies should not be accepted uncritically as accurate representations of the degree of relatedness between organisms."[245]

In 2003 Conant and Wagner of the University of New Mexico concluded that whole genetic networks in *E. coli* and *S. cerevisiae* were the product of convergent evolution.[246]

Vowles and Amos of Cambridge University in 2004 concluded that the sequences surrounding microsatellites show "conservation over unexpectedly large tracts of evolutionary time... we also find evidence of convergent evolution,"[247] applying to DNA segments that make up approximately 30% of the human genome that was previously deemed 'junk'.

Then as genome sequencing became cheaper and easier to analyze, between 2010 and 2015 the following examples of genetic convergence began to appear.

- Maeso, Roy, and Irimia of the University of Oxford concluded that the recent surge in genome sequencing has "unveiled an unexpected wealth of cases of recurrent evolution of strikingly similar genomic features in different lineages".[248]
- Erich Jarvis of Duke University found "many similar genetic changes" in the sequences that regulate speech in the brains of humans and singing songbirds, such as zebra finches.[249]

- A team found a common mutation for pigmentation among divergent species such as bananaquits, Japanese quail, and chickens; another common mutation among Monarch flycatchers, domestic pigs, and several strains of sheep; and another among arctic skuas and pocket mice. "In all of these cases, because the species sharing a common mutation are so divergent, mutational convergence probably represents the independent origin and subsequent selection of the same mutation."[250]
- Christin, Weinreich, and Besnard of Brown University concluded that an accumulation of studies showing convergent molecular evolution ranging from "gene families being recurrently recruited to identical amino acid replacements" have surprised researchers and caused a rethink about the possible outcomes of genome evolution.[251]
- Echolocation (sonar) has developed at least twice in bats and in toothed whales. When genome-wide analysis was done on the three lineages, researchers found "extensive convergent changes" in the genetic basis to echolocation. The changes show that molecular convergence is "not a rare process" and is "widespread, continuously distributed and commonly driven by natural selection".[252]
- Sticklebacks (a type of fish) have adapted to live in fresh water all over the world, and researchers discovered that the genomic changes behind the adaptation are always the same.[253] David Marques of the University of Victoria said that the find supports "the emerging view in evolutionary biology that mechanisms underlying adaptive evolution are often highly repeatable and thus may be predictable... with a limited set of tools [that] can lead to convergent 'solutions' to common problems".[254]
- At Max Planck Institute, mouse populations in the wild

that developed extremely large bodies were compared against artificially selected heavy mice. Both changed genetically in similar ways, suggesting that, "artificial selection in the laboratory changes the same loci in the genome as natural selection".[255]

- Among 18 insects in four orders surveyed in yet another study at Cornell University, all developed resistance to a certain powerful plant toxin by the exact same gene mutation, and a further 11 of them carried the same second mutation to the gene. A researcher noted, "evolving resistance to the plant toxin had very few effective options... there has been tremendous repeatability, even at the molecular level".[256]

- In a survey of 29 distantly related insects by Princeton University, 14 species showed "nearly identical" genetic characteristics for adapting to a plant toxin. Senior researcher Peter Andolfatto said, "Is evolution predictable? To a surprising extent the answer is yes... The fact that many of these solutions are used over and over again by completely unrelated species suggests that the evolutionary path is repeatable and predictable."[257]

- Olsen, Kooyers, and Small of Washington University in St Louis found convergent gene deletions, not just nucleotide changes, happening repeatedly in multiple species of clover. Olsen shared, "[This] gets at the question of how constrained evolution is. The more constrained, the more predictable it is..."[258]

- Charles Darwin mentioned electric organs in fish as an example of convergent evolution. In 2014 a team of researchers were shocked to find that the genetic tools and pathways behind all lineages of electric fish were the same, and had arisen independently at least six times.[259]

- When researchers at UC Santa Barbara sequenced all the genes related to bioluminescence in squids, they were enlightened by the convergent expression of thousands of genes. In animals that evolved light organs independently, the gene expression profiles were strikingly similar, so much so that the researchers were able to predict the results.[260]

- Buller and Townsend of Johns Hopkins University studied protease (enzymes that cleave proteins) and found relationships that constitute "incontrovertible links between structure and function and limit the fitness landscape" of the enzymes. Among two classes of protease, "each class is subject to unique structural constraints that have governed the convergent evolution of enzyme structure."[261]

Implications

To the issue of whether such striking examples of genetic convergence should bring into question assumptions of relatedness, Christin, Besnard, and others went on to write,

> ... in some realistic conditions, even a relatively small proportion of convergent codons can strongly bias phylogenetic reconstruction, especially when amino acid sequences are used as characters... we recommend [methods] to identify potential phylogenetic biases and avoid evolutionarily misleading conclusions.[262]

Similarly, on the role that convergence has and should play in genetic studies, Parker and others wrote,

This study represents the first systematic attempt to provide a framework for the genomic analysis of sequence convergence associated with independently shared phenotypes. Our findings strongly suggest that, despite many recent papers reporting sequence convergence in particular candidate genes, the importance of this mode of molecular evolutionary change is relatively underappreciated, and is under-exploited in seeking to understand the genetic basis of complex traits such as echolocation.[263]

Researchers at the Department of Biochemistry and Molecular Genetics; University of Colorado School of Medicine found a remarkable case of amino acid convergence in mitochondrial genes in snakes and lizards. They write,

Just a few of these convergent substitutions were sufficient to positively mislead the inference of phylogeny, even with thousands of sites providing latent support for the correct underlying relationships. Since this example demonstrates that molecular convergence can happen en masse in nature, affecting multiple genes, it is important to consider the threat this poses to molecular systematics, and careful genome-wide assays for convergent molecular evolution are warranted. This result implies that the protein adaptive landscape is sometimes highly constrained.[264]

More genetic convergence

In the years since 2015, numerous academic papers have tried to analyze the role of molecular convergence, how to categorize it, and how to recognize false positives. The research is being propelled by an incredible amount of newly available data. Draft and fully sequenced genomes were rare in the 1990s, but by 2010 the trickle turned into a hose, and now a wave of data. People have their eye on sequencing every genus of vertebrates, all 10,000 of them, and some extinct ones.

We are in the early stages of analyzing the data. People are coming to a variety of conclusions, some supportive of molecular convergence and some not.

For example, a group of researchers sequenced the genomes of three mammals that independently became aquatic: killer whale, walrus, and manatee. They compared these with the available genomes of dolphins and other mammals to see if the convergent phenotypes are driven by convergent genetic changes. They did find "convergent molecular evolution among the marine mammals." However, they found more convergence when comparing to sister lineages, such as manatee to elephant. So they concluded that phenotypic convergence rarely drives genetic convergence.[265]

A separate analysis of the same data found "pervasive convergence at the gene level during mammalian shifts to the marine environment" across "hundreds of genes" related to aquatic adaptation. They explain that the earlier study,

> employed the hypothesis that convergence would be seen as single amino acid sites changing to the same amino acid. They did not find an excess of such changes over background... As an alternative approach, we introduced a novel analytic framework that focuses on those genes showing convergent acceleration or

deceleration in their whole-gene evolutionary rates. This strategy represents a more course-grained perspective to identify common sets of genes and pathways that experienced a shift in their evolutionary pressures during phenotypic convergence. Our application of this gene-centric strategy to marine mammals uncovered a strong statistical excess of convergent shifts in evolutionary rate. This suggests that convergent changes in selection at the gene level is relatively common compared to convergence at single amino acids, and that its signal is more readily detected. These two approaches operated at distinct levels but employed the same basic data set.[266]

Other studies have similarly produced a mix of convergence and variability, but there are some notable additions on the side of genetic convergence.

In 2017 a similar molecular study found "strong evidence of convergence" among mammals adapting to life underground. They used a similar technique by looking at evolutionary pressure on genes related to skin and vision, and they found convergence among "several uncharacterized genes and regulatory sequences".[267]

The giant panda and red panda both convergently evolved to eat bamboo from meat-eating progenitors. The genetic basis for the changes was not known until 2017 when a team of Chinese scientists sequenced the red panda genome and improved on the giant panda's, then performed a whole-genome analysis to see if there was genetic convergence to mirror the phenotypic kind. They found that "genetic convergence occurred at multiple levels spanning metabolic pathways, amino acid convergence, and pseudogenization... analyses revealed adaptively convergent genes potentially involved with pseudothumb development and essential bamboo nutrient utilization", as well as taste receptors.[268]

The genes of the two largest groups of fungi (Ascomycota and Basidiomycota) were analyzed and a draft review in 2019 showed that "despite >650 million years of divergence, the same genes have repeatedly been co-opted for the development of complex multicellularity... resulting in >81% convergence across shared multicellularity-related families." Interestingly, they found that many shared genes used for multicellularity predate multicellularity,

> suggesting that the coding capacity of early fungal genomes was well suited for the repeated evolution of complex multicellularity. Our work suggests that evolution may be predictable not only when organisms are closely related or are under similar selection pressures, but also if the genome biases the potential evolutionary trajectories organisms can take, even across large phylogenetic distances.[269]

We are seeing the same unfolding of analysis with respect to genomes that we saw 20 years earlier with respect to phenotypes: surprising levels of convergence, strong selective pressure, and evolutionary constraint causing conserved genetic sequences over surprisingly long periods of time. With the fungi examples we also return to independent multicellularity and an emphasis on process and repeatability.

Initial comparisons of thylacine and wolf – remarkably similar marsupial and mammal predators – in 2018 showed little evidence of molecular similarities. However, another study in 2019 compared their non-coding elements and found that natural selection independently made the same genetic changes to form similar bone, cartilage, and muscles of the face, as well as convergence in regulatory elements of the brain.[270]

Genetic constraint

Based on standard theory, the generous presumption has been on the side of similar genotype coming from common ancestry. That is mostly true. 70% of human genes are shared with the acorn worm and are thought to be conserved at least 550 million years.[271] If several lineages of vertebrates came from independent multicellularity events, they would share a patchwork from the common genetic toolkit. This is in a sense common ancestry, but the tree model begins to break down the closer you get to the trunk and roots. It is more relevant to discuss commonalities from the gene-sharing root system and common ancestry among a group of similar forms that most likely sprung from their own trunk.

This idea of relatedness must be coupled with an understanding of independent multicellularity, horizontal gene transfer, high constraint from common cellular origins, and of course, convergence.

For example, 50% of banana genes are found in humans,[272] but it has already been shown that plant life came from its own unique eukaryotic origin. Animals did not evolve from plants and they have no common ancestor (See Ch. 4: Endosymbiosis). How should we refer to the shared genetics? The language has not caught up yet, and many studies are referring to them as conserved from an ancient common ancestor. 'Common ancestor' excludes the role of gene-sharing in primitive cells. 'Conserved' implies that the novelties only happened once and excludes the role of repeatable convergence. 'Constrained' seems better as it implies limited options, but it still seems to lack common origin. Perhaps constrained from common gene-sharing would be a better description, but even this lacks the role of viral exchange that can happen throughout the history of a species (see later this chapter).

Protein sequence space

The prevalent view is that DNA similarity is like language similarity. When Europeans landed on the Hawaiian Islands for the first time, the natives were speaking a language that largely resembled a Polynesian dialect. This was irrefutable evidence of colonization from or contact with Polynesian islands, because language has so much possible variability that no rational person would believe that two independent cultures would converge on mutually intelligible languages. It doesn't matter whether a coconut is called 'ng niyog', 'kelapa', or 'kokoc'. All that matters is that the person being spoken to understands that when those sounds are made, it's referring to that one thing. There are common etymological sources because there is a finite set of easy sounds to make, and duration and complexity need to be considered, but in general and with few exceptions, there is no selective advantage to the pronunciation of a word.

The comparison of species divergence to language divergence was used originally by Darwin and further mentioned as recently as Jerry Coyne's *Why Evolution is True* (2009). This is the common view of DNA, that it is so accidental that similarities must imply common origin. This is not the case, however, because DNA is not random and arbitrary. It does matter how a protein is described. There are limited ways to pronounce 'coconut' in the genetic code. Genotype leads to phenotype. The environment and phenotype drive genomic changes. DNA variation is the mechanism for changes leading to natural selection and it has the potential to provide a strong selective advantage. It is highly constrained where language is not.

On a fundamental level all human languages share commonalities in pronunciation, structure, or the way symbols are used, just as all biological organisms on earth share fundamental attributes. Of the 64 possible codons, the combination CGU always codes for the amino acid arginine, as does CGC, CGA, CGG, AGA, and AGG. The genetic code is universal, the same code makes the same amino acid in all known life, but within what is fixed there is some variability, with six ways to code for the same thing in the case of arginine. There is little variability to the sequence of amino acids that make a particular protein, yet due to variability in the genetic code, there are different ways of writing the code for the exact same sequence of amino acids. If the same protein evolved independently in different species, it would appear with the same amino acids and within the limited codes to describe the protein. Like code through a software compiler, the code to write useful proteins is limited.

How many useful proteins exist? Certainly the repertoire is quite extensive, but ultimately there is a biological solution set available that is finite, and in a particular environment the solution set would narrow considerably, making convergence a strong possibility. To this end, faculty at the University of Edinburgh published their conclusions in the *Journal of the Royal Society* that "protein sequence space" has been quite plausibly entirely explored in the history of life on earth.[273] They write that there are two general assumptions made when considering molecular evolution:

1. The size of protein sequence space, i.e. the number of possible amino acid sequences, is astronomically large.
2. Only an infinitesimally small portion has been explored during the course of life on earth.

According to the authors, the first assumption has been shown incorrect numerous times, but they go on to dispute the second by considering "the number of organisms, genome size, mutation rate and the number of functionally distinct classes of amino acids". Their calculation shows "no support for the idea of contingency at a molecular level and it provides strong support for the ideas of convergence".

If one were to rerun the tape of life, as Gould suggested in *Wonderful Life*, we would probably still see coconuts, and their protein composition would be quite similar. Gould's question can be phrased another way that is actually testable: If another planet just like earth developed life, would the morphology and protein composition be similar to those of earth? While scientists don't predict the future, the constraint observed on earth should generate a resounding, "That is a logical conclusion!" from the scientific community.

The 99%

While discussing my book, I have sometimes heard, 'But human and chimpanzees are 98% similar genetically?' I have responded, 'What percentage would you expect for separate ancestry?' To which I get a confused look.

The use of a percentage similarity is slightly deceptive and needs context to even make sense of it. Take for example the much quoted 99 or 98% similarity between chimps and humans, later revised to 94%. The actual percentage similarity can vary considerably depending on what is being compared, the lowest appearing at 87% similarity.[274] The chimpanzee genome is about 11% larger than the human genome, so a comparison resulting in 99% similarity is drastically over-simplified.

Many comparisons only use protein-coding genes (a small percentage of overall genome), and not even all of them. Many quoted percentage similarities don't distinguish between comparing base pairs or amino acids or proteins, whether from mitochondrial or nuclear DNA, all of which would produce differing results. Are the studies including the regulatory sequences outside of genes? Are they including the large sections of repeated code that appear to be structural? How do you incorporate the same gene copied several times? In fact, 6% of the 25,000 human genes have no analogues anywhere in the chimp, but 80% are different by at least one amino acid.[275]

Expand the view and you get some strange results. Humans share 50% of genes with bananas, 70% with brainless worms, 75% with chickens, and 94% with chimpanzees. Would two unrelated creatures never use the same protein? Would two unrelated creatures have 0% similarity? How similar would two similar forms from unique multicellular origin be?

Just comparing two strings of randomized DNA characters will show 25% similarity in base pairs, but even this is misleading, because it implies that DNA is arbitrary and totally random.

Genes use three letters to code for an amino acid, leaving 64 possible combinations describing 20 amino acids and the start/stop codons. Thus, if two organisms independently evolved the same amino acid, the two codes would have about a 79% similarity in base pairs.* So as two species diverge, a constrained gene from a common ancestor would move from 100% towards 79% similarity given enough time, ignoring several assumptions. But if the gene were evolved independently in separate ancestry, the result would also average 79% similarity when comparing base pairs.

However, comparisons at the protein level should have higher levels of similarity among a wide variety of life, and they do. This is a reflection of the limited protein space, high constraint from common gene-sharing cells, and convergence driven by natural selection. A high degree of protein similarity (>70%) has already been demonstrated for widely different forms without common lineage, and similarity should only increase between similar forms because there is a causal link between genotype and phenotype.

How does a percentage similarity contribute to a discussion

---

* 1*23/64+(2/3)*(64-23)/64 =0.786

on evolutionary relationships? In 2000 Jonathan Marks wrote in the 'Chronicle of Higher Education' later reprinted in a textbook of anthropology,

> Using the figure [of 98% genetic similarity] ignores the context necessary to make sense of it... the genetic comparison is misleading because it ignores qualitative differences among genomes. Genetic evolution involves much more than simply replacing one base with another. Thus, even among such close relatives as human and chimpanzee, we find that the chimp's genome is estimated to be about 10 percent larger than the human's;... and that the tips of each chimpanzee chromosome contain a DNA sequence that is not present in humans.[276]

Some differences are structural. "Chunks of missing DNA, extra genes, altered connections in gene networks, and the very structure of chromosomes confound any quantification... Entire genes are also routinely and randomly duplicated or lost," according to evolutionary biologist Jon Cohen.[277] He quotes a further 2006 study that showed a 6.4% difference in gene copy numbers between humans and chimpanzees, concluding that "gene duplication and loss" plays a greater role than base pair substitutions, which are the subject of the widely used comparisons.

Chromosome fusion

The global view of the chromosome is quite complex. Each chromosome has telomeres at the end, representing long chains of repeating code, whose function is to provide stability, like the plastic on the end of a shoelace. Since the telomeres can break apart and be lost, there is a declining scale of importance to genetic material as it moves farther away from the centre, similar to how the most disposable organs are farthest from an animal's body core. Even the location within the chromosome plays a huge role in how traits vary and evolve,[278] and highly conserved sequences tend to be closer to the middle where they won't be lost. The loss of excess genetic material from the ends then provides chromosomes a way to lose non-essential elements and become leaner. Changes to the number and structure of chromosomes can therefore add to overall fitness and be selected for.

All apes have 48 chromosomes (24 pairs), but humans have 46 (23 pairs). It was long theorized that if humans and apes share a common ancestor, then there will be a human chromosome that is the fusion of two chromosomes into one, so a human chromosome with telomeres in the middle would be found. The human genome project found something similar to a fusion in human chromosome number two, and Ken Miller presented it as evidence supporting evolution during the 2005 Dover case regarding Intelligent Design being taught in schools. The fused human chromosome has general similarities to two separate ape chromosomes, as well as some differences.

Chromosome quantity is not the most relevant piece to genealogy. Almost the opposite, the first mapping of a large number of genomes found that the number and size of chromosomes are correlated across all species and follow an S curve when mapped, implying that there are undiscovered laws governing chromosome structure that all genomes adhere to.[279] This unknown pressure drives chromosome splitting and recombination. The function of these structural modifications is poorly understood.

A mosquito has 6 chromosomes, a giraffe has 62, a type of fern has 184, and a tobacco plant has 48. As chromosome shaping is an inherent part of standard evolutionary theory, the telomeres in the middle of a human chromosome indicate that an inversion or fusion happened, but not necessarily that it came from two chromosomes of a common ancestor with apes.

The logic is reminiscent of 'Abdu'l-Baha's response to evidence for trace members as supporting the theory of humans evolving from animals. It is a fact that humans have remnants of what used to be a tail, indicating that primitive humans had tails. It is a fact that living monkeys have tails. This does not prove that humans and monkeys have a common ancestor. Likewise, it is a fact that humans have genetic traces of a chromosome fusion, indicating that primitive humans likely had 48 chromosomes. Living apes have 48 chromosomes. This does not prove that humans and apes have a common ancestor. The doctrine of common descent requires there to have been a fusion, but the presence of a fusion does not falsify the theory of independent descent because there is a yet-to-be-elaborated function in the chromosome fusion.

Endogenous retroviruses

Viruses are not much more than strands of DNA. In fact, much debate has gone on about whether they should be classified as 'life' or not. Retroviruses hijack the machinery of cells by inserting their own into the cell's DNA, and then the cell structure starts replicating the DNA of the virus. The immune system fights these hostage situations with a complex method of marking the infected cells for destruction. If a virus inserts itself into the body's cells and is not killed, it will remain in the DNA. If the cell being attacked is an egg or sperm, then the virus may be duplicated into future generations and is called endogenous.

The human genome has thousands of such viruses (at least 8% of genome[280]), so when the human genome project and the chimpanzee genome project compared notes, there were families of endogenous retroviruses (ERVs) in analogous locations in the DNA.

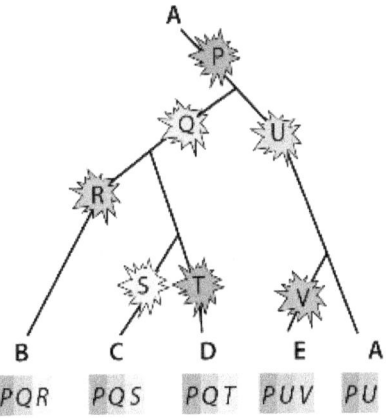

Example model of ERV relationships

The theory behind these ERVs goes something like this: an ancestral species evolves and diverges in a branching pattern that results in five distinct species. Along the path of evolution, and along different branches, each species gets hit with endogenous retroviruses. Each time a virus infects a species, the viral DNA gets passed down to all its descendants, but the virus will not appear on any sister lineages. So if the original ancestral species were infected with a virus, then all descendants would have it.

This is what one would expect, and this is what appears: in the same locations in the genomes, some viruses are unique to primates, some are shared with humans and mice, some with all mammals, some only to birds, and others are found in some human populations, but not all humans. Case closed on the common ancestry of all mammals including humans.

Maybe not. The given scenario is correct, but incomplete. It is presumed that ERVs have no value to the host; that ERV insertions are in random locations; that ERVs do not infect multiple species at the same time; and that the virus code found in the genome was inserted during an attack.

Thanks, virus

What's the most successful virus? One that does not kill the host. If a virus causes paralysis, neuropathy, or respiratory failure, it will be difficult to propagate the virus over time through human DNA. Even better, if the virus provides a selective advantage to the host, the virus is more likely to exist and be passed along in sperm and egg cells.

For decades experiments have tested genetic value by knocking out sequences of code and trying to grow the resulting embryo. This method can indicate function. If the embryo failed to develop properly, then the sequence was necessary.

In the cases of ERVs, these knockout tests have consistently surprised researchers. In 2006, Texas A&M and University of Glasgow showed that a class of retroviruses were critical for the placenta development and immune response in sheep. They concluded, "This infection was beneficial to the host and was then positively selected for during evolution... these days, sheep cannot do without them."[281] There are two retroviruses acquired independently in both mice and humans that were knocked out of mice in a 2009 experiment, resulting in their death in utero at about 12 days of gestation.[282] Another 2009 experiment surprised researchers when they realized that both rodents and humans use ERVs to control the expression of the same genes, concluding that "not all retroviral remnants in our genome are simply junk DNA".[283] Some viruses in the genomes of fish, birds, and mammals can selectively kill tumor cells, and may have been selected for their role in controlling cancer.[284] A 2011 study identified 1,500 viral genes used in the uterus of mammals that were previously called junk.[285] In 2015 researchers found 13 RNA sequences necessary for a fertilized human egg to develop that were derived from ancient viral infections.[286]

Evidence also suggests that the genome holds onto certain viruses in a suppressed, controlled state, possibly as a permanent record for the immune system. With a few exceptions, all of the existing human ERVs are defective as viruses.[287] A genetic switch has been found, which when removed in mice allows a variety of the viral DNA to "come alive" and begin destroying the cell with detrimental changes.[288] The reality of ERV insertions is not as clean cut as was first described. There are ERVs found in chimpanzees and gorillas that are absent from humans and orangutans[289] (orangs supposedly split before chimps and gorillas), and there are others that show the same ERV with different insertion locations between chimps and humans.[290] Another ERV is found in all baboons, but is also found in some species of cats, implying a horizontal transfer of virus, or that baboons evolved from a few species of cats exclusively.[291]

When a virus inserts its DNA into a cell, that cell will carry the viral DNA in a particular location in the genome. The evidence for common ancestry stems from an assumption that the insertion site is random. Thus, two genomes with the same virus in the same location must have derived from a common ancestor who was infected. But retroviruses (and retrotransposable elements) of all kinds show preference for genomic insertion sites,[292] implying that the insertion sites are not random.

Some sites may be more suited or vulnerable to a particular virus becoming endogenous, or other infection sites may be unsuitable. Viruses generally prefer certain types of regions in the genome, so there may be random attachment within the same region. Natural selection also applies to viral infections. Less suitable attachment points will be weeded out, and those that add to fitness will be propagated. If viruses prefer certain regions of the genome (or certain loci in some cases), and natural selection favors particular attachment sites, then retroviruses will appear in the same locations among the genomes of multiple species, but from parallel infection.

HIV jumped from apes to humans, and it has a distinct DNA target site preference, as does avian sarcoma-leukosis, which infects mostly chickens but also some mammals, and murine leukemia, which infects mice and several other vertebrates.[293] The populations that would be susceptible to inter-species horizontal infection are those that have similar immune systems, such as apes and humans. To then argue that the similarity of immune system was derived from a common ancestor enters a circular argument: if a derived similar immune system is evidence of common descent, then it explains many ERVs, in which case the ERVs themselves are not definitive evidence of common descent. And by the way, going by immune system similarity, humans are closest to gorillas, not chimpanzees.[294]

All this aside, viruses are a form of horizontal gene transfer, some of which transferred among the common root system of singular cells. If the majority of genes are conserved from primitive gene-sharing cells, most likely the majority of viral DNA are also remnants from before multicellularity was a reality, and thus before the species threshold was crossed.

Virolution

While distinct retroviruses make up at least 8% of the human genome, about 42% is made up of fragmented ERV derivatives, such as repeated sequences.[295] That means that a significant chunk of the human genome comes from a form of horizontal gene transfer, data not derived from descent with modification.

A study published in 2015 took advantage of the large volume of high quality genomes across 26 animal species to compare transcribed genes and try to catch instances of horizontal gene transfer.

It found 33 new examples of transferred genes unique to humans, bringing the total to at least 145 genes that were not derived from ancestors and transferred in an advanced stage of evolution. Still a tiny fraction, but there could easily be many more waiting to be found. It also found, on average, 173 transferred genes in worms, 40 in flies, and 109 in primates.[296]

Such levels of genetic exchange in highly evolved animals is still not widely acknowledged. A paper in 2019 found that a package of iron-scavenging genes went from bacteria to fungi, conferring a complex new trait. The author noted that despite the iron-clad case for the transfer to a eukaryote, there are "still people who are staunchly opposed to it as a phenomenon".[297]

Overall similarity might not just be attributed to primitive gene transfer and convergent evolution, but also more recent gene transfer in advanced stages of evolution.

For example, a sea slug off of North America's east coast was recently discovered to use photosynthesis, just like a plant.[298] Dozens of genes used in algae to process solar energy have been identified in the slug, believed to be the result of horizontal gene transfer. Similarly, a eukaryotic algae (*Galdieria*) thrives in toxic environments thanks to genes that it borrowed from bacteria.[299] In other cases, gonorrhea – a disease exclusive to humans – was found with sequences of DNA identical to those of humans.[300] A Malaysian parasite plant was found with 49 genes from its grape host and is actually using three-quarters of the heisted DNA.[301] In another case, a cluster of 23 genes appeared to have moved from one strain of fungi to another.[302] Plants everywhere can horizontally share their entire chloroplast genome, which are often most similar among species growing in the same area.[303]

Perhaps the most dramatic example of co-evolution with viruses is the yeast *Scheffersomyces stipites*, which uses a modified genetic code. In other words, it does not use exactly the same code for amino acids that most earthly creatures use. There are many such organisms, but researchers found a virus with the same modified genetic code attacking the yeast! What's more, the virus managed to leave viral DNA with the host, not once, but at least four times in its evolution; and the host ended up using a viral gene to make a protein for its own benefit![304]

For the process of natural selection, there must be variation to select among. The emphasis in traditional evolutionary theory has been on point mutations and genetic inheritance, and gradualists have been increasingly assaulted over the inability of these processes to explain observable facts. Viral DNA delivers entire sets of new genes all at once, rather than requiring the gradual point mutation emphasized by Darwin, Dawkins, and others. The full force of the discovery is only slowly hitting mainstream science. In 2009, Frank Ryan published *Virolution*, with the slightly presumptuous subtitle: 'The most important evolution book since Dawkins' *Selfish Gene*.'[305] According to Ryan, the first revolution came with the enlightenment in the 1990s of epigenetics – heritable traits that are not derived from DNA; the second revolution came with the enlightenment of hybridization, its prevalence and dramatic ability to form new karyotypes; the third revolution (or shall I say, virolution?) came with the 2001 completion of the draft human genome and the gradual realization that the genome was simpler than imagined and made up of large indispensable viral elements.

Ryan's book is fascinating and ultimately proposes a model of viral symbiosis with great examples that elaborates a new and novel understanding of evolution. It shows that viral DNA is much more prevalent and plays a greater role in evolution than imagined. A better understanding of viral DNA also refutes assumptions about DNA only changing gradually with point mutations.

Until recently common ancestry and descent with modification was the only tool in the biologist's kit to explain similar genetic elements. Thanks to recent elucidations, the kit also includes convergence, viral DNA, and constrained growth from a common root system.

In addition to Ryan's list of genetic revolutions (epigenetics, hybridization, and viral DNA), below are four other revolutions in DNA analysis that all appeared at about the same time (2009–2011) and are all similarly revolutionary in terms of their novelty and dramatic effect on DNA analysis. All these revolutions, taken together, show that humanity's collective understanding of genetics is still in its infancy; a new long-term conceptual framework is developing for evolutionary biology.

1: DNA topography

A new field of research is opening up a radical approach to DNA analysis. The traditional approach reads the four-letter alphabet and records the sequence of letters. However, those letters are assembled in a three-dimensional structure on the chromosome, and the 'topography' of the structure forms a lock-and-key fit with proteins that influence DNA expression.

Thomas Tullius of Boston University spent more than 20 years studying the structure of the human genome. He joined a team of other researchers to develop a method "for uncovering functional areas of the human genome by studying DNA's three-dimensional structure".[306] Their paper, published in a 2009 journal of *Science*, indicates that,

> Topography-informed constrained regions correlated with functional noncoding elements, including enhancers, better than did regions identified solely on the basis of nucleotide sequence. These results support the idea that the molecular shape of DNA is under selection and can identify evolutionary history.[307]

Very similar DNA sequences may have entirely different topographical shapes, and dissimilar DNA sequences may have the same topographical shapes. The study compared several species with humans and found that the topography-based analysis found twice as much evolutionary constraint as compared to the sequence-based approach. If the analysis is correct, the study has identified several new functional areas of the genome.

This analytical tool has opened up an entire new angle to studying genome function and evolutionary relationships. The study found many new functional areas in the non-coding part of the genome, further emqhasizing the fallacy of calling it 'junk'. This technigue should be used in any serious analysis of evolutionary relationships.

2: Ps and Qs

The lowercase letters 'p', 'g' and 'q' show a slight resemblance and could easily be confused for each other if the reader is not looking closely (as happened twice in the preceding paragraph).

A similar problem has been discovered in the DNA code. Cytosine, one of the four bases, can have a small extra molecule attached. When cytosine gets the freeloading methyl group it becomes 5-methylcytosine, and it is (sort of) a fifth character in the DNA code. This has been known since the 1980s and plays an important role in passing down temporarily heritable traits. Peeking into the world of 5-methylcytosine has opened up the field of epigenetics and provided an entirely new view of diseases, disorders, and evolution.

In 2009, and again in 2015, scientists stumbled upon previously undetected methylated bases in eukaryote cells: 5-hydroxymethylcytosine[308] and methyladenine.[309] These contributors to the epigenome can now be read and studied. Now that the unsuspected tools to differentiate 'p', 'g' and 'q' are available, the book of the genome can be re-read to determine what role they play.

3: Natural genetic engineering

James Shapiro's *Evolution: A View from the 21st Century* (2011) documents a new addition to evolutionary theory. He describes standard theory as moving from "descent with modification" and "natural selection" to an increasing number of interconnected hypotheses, such as three domains of life, evolution of development, genomic networks, symbiogenesis, horizontal gene transfer, epigenetics, hybridization, descent with modification, and natural selection. Adding to this list, he articulates "natural genetic engineering".

According to classical theory, the cellular genome is a Read Only Memory (ROM) that only changes from copy-error accidents. Six decades of research has replaced the ROM view with a read-write storage system that is being intentionally molded and changed by the cell. One such cellular function is the ability of immune system cells to reformat their genome to store the DNA information of antibodies to viruses. Without such ability, we would not have an adaptive immune system.

Other cellular functions actively prepare genome sites for cut-and-paste or copy-and-paste, then prepare the new location to paste it. These transposons (transposable elements) were traditionally thought to be random, similar to viral DNA, but now they're known to be largely an intentional process of the cell. These cleavage and attachment sites provide "one well-established mechanism for major chromosome rearrangements observed in the course of evolution".[310]

Instead of the cell being controlled by "an all-determining genome, we now understand how cells regulate... their DNA molecules".[311] The cell formats the rewritable USB 'flesh' drive as it sees fit for easy access to information, and this reformatting is part of the normal life cycle of the cell. Sometimes the cell even doubles its genome just to expand hard drive space and have more room to work. The genome is a tool for the cell, not the other way around (Dawkins' next best-seller: *The Selfish Cell*).

According to Shapiro, the view of evolution from the 21st century implies that

> [I]t is apparent that systems engineering is a better metaphor for the evolutionary process than the conventional view of evolution as a selection-biased random walk through the limitless space of possible DNA configurations.[312]

Mobile genetic elements have traditionally been viewed as random mutations that accumulate 'junk' DNA. Shapiro's work, among others, suggests that the cellular genome is a read-write memory that is subject to nonrandom rearrangement. It also suggests that in the drama of evolution, natural genetic engineering is the star of the show and descent with modification is the backstage help.

Viewed properly, retrotransposable events are part of a purposeful process that will result in convergence.

4: Genetic predisposition

In 2001 and 2008, Christian Schwabe proposed that evolution is defined by species repeatedly appearing in embryonic form with a distinct mature potential (see Ch. 4: Independent birth and genomic potential). He also made the prediction that if the theory is correct, genome analysis will reveal immature forms with genomic potential of some higher life form.

Around the year 2010 large amounts of data began to accumulate beyond the standard model organisms (fly, mouse, yeast) that were the staple of genetic research in the 20th century. Scientists went head first into the world of big data, where the scientific method, with its hypotheses and experiments, became less relevant. When sifting through a mountain of data for relevance, the focus is on signal from noise, exceptions, pattern recognition, and cyclical behavior.

The new sets of data began to unravel many assumptions about relationships between fossils. For example, there is a common mussel found in the Atlantic on the coast of Sweden that is identical to one on the Pacific coast of North America; however, they are genetically different.[313] A more familiar example would be giraffes, thought to be a single species until DNA analysis in 2008 revealed at least six distinct lineages that have not interbred for 0.1–1.6 million years.[314] Similar stories can be found in thousands of examples of frogs, beetles, fish, butterflies, fungi, and worms. In such cases, it is not obvious whether the similar forms appeared through divergence from a common ancestor (as in the giraffes), or convergence from separate ancestry (as in the mussels).

The name for these similar-looking but genetically-different creatures have a fun biological term: cryptic species. DNA analyses of worms have provided the greatest breadth of cryptics. Worms previously lumped together due to common appearances have been identified as having genetic traits that put them on wholly different paths of evolution.

Some primitive forms have been found with the genetic foundation for some distinct higher order of life. Since 2008, the following discoveries have been made:

- A newly devised technique to analyze DNA discovered a counterpart to the cerebral cortex[315] and spinal chord[316] in an invertebrate marine ragworm (*Platynereis* spp.). This worm, then, has the primitive foundation that could lead to a brain.

- Other tiny marine worms, such as *Xenoturbella* spp. and *Acoelomorpha* spp., have genetic similarities to vertebrates, yet they lack a developed nervous system or gut.[317] They have been described as evolving backwards into more simple forms from a more complex common ancestor.

- The brainless acorn worm *Saccoglossus kowalevskii* broke off from the evolutionary tree before the emergence of vertebrates, yet has genetic sequences that regulate vertebrate brain development.[318] This too has been described as having evolved backwards from a brainful progenitor.

- *Xyloplax* spp. was once thought to represent a new class of animals, until DNA analysis showed that it is a primitive form of starfish, or as Daniel Janies of Ohio State University concluded, "many could not recognize it as a starfish until we unlocked its genome and development".[319]

- *Nematostella vectensis*, a brainless sea anemone, uses the same genes that control head development in higher animals.[320]

- The jawless sea lamprey *Petromyzon marinus* has gene expression similar to jawed animals.[321]

Notospermus geniculatus (top) and Phoronis australis (bottom).

- *Notospermus geniculatus*, a ribbon worm, and *Phoronis australis*, a horseshoe worm, look very different (see photos), but genome analysis found that they are evolutionarily related. They also share many gene families and arrangements with vertebrates, such as head development.[322]
- Common marine sponges, among the most primitive of animals, carry a toolkit of sophisticated genes used in eyes, brains, and central nervous systems. "As it turns out, sponges have the genetic potential that could lead to

things like eyes, legs and arms," says April Hill, professor of biology at the University of Richmond.[323]

- The two largest groups of fungi (Ascomycota and Basidiomycota) co-opted the same genes for multicellularity. Reviewers decided that "the genome biases the potential evolutionary trajectories organisms can take."[324]

Was Schwabe right? Did these primitive creatures evolve backwards into simpler forms, or are they the herald of a new vision of evolutionary biology? Time will tell. The technology to distinguish such capacity is just a few years old, and nobody, not even Schwabe, has taken up a review of their significance as a whole. Further scientific inquiry may identify these worms as being in the early phase of a mature species.

These discoveries are very difficult to fit into the current evolutionary paradigm, and they represent strong evidence for genomic potential in primitive forms.

If an independent multicellular origin is established for numerous groups of species, all the way to apes, it will be an astonishing paradigm shift that will support independent descent, but even such an amazing validation would still conflict with my understanding of 'Abdu'l-Baha. He spoke of the potential in an embryo and suggested that the unevolved progenitor of humans was distinctly human in potential. Science could easily validate a separate lineage for humans and still conflict with the idea that there was ever a tiny organism that was distinctly human in potential.

CHAPTER 10

HUMAN POTENTIAL

Tiktaalik

The age of the Internet and the exponential increase in cheap processing power has revolutionized how data is gathered, analyzed, and shared. There have been significant discoveries since 2000 shifting scientific thought about how creatures evolve. A new long-term conceptual framework is developing; it can be seen in many authors writing about the implications of convergence, experiments in evolution, hybridization, constraint, horizontal gene transfer, the role of viral DNA, natural genetic engineering, and DNA comparisons. Many have attempted to bring to prominence some new revolutionary idea, but few have put everything together into an overall framework, and despite the pleas of leading researchers, the ideas of gradualism and universal common ancestry still permeate science literature.

What discoveries might happen in the next few decades? Darwinian evolution theorized a transitional fossil at a certain age that would represent a fish moving toward becoming a quadruped land animal. Such a fossil was discovered in 2004, named Tiktaalik, and has been presented as one of many finds that validate the view of single origin for all life (although fish have evolved the ability to live on land many times).

The multiple-tree model that I've described is a scientifically plausible approach to evolution that is consonant with the apparent meaning of 'Abdu'l-Baha on human origin. I've already hypothesized several areas of investigation that will bolster the theory: comparing attributes of earthly and extraterrestrial life; discovering the processes behind the formation of simple cells, eukaryotic cells, and multicellularity; confirming several proposed unique origins within the vertebrate lineage; identifying function throughout the human genome; uncovering distinctly human fossils that lengthen the perceived divergence time with apes; identifying numerous creatures with genetic traits of specific more complex life. An important addition to this list will be finding human potential in primitive forms.

Genetic species

One might conclude that the line that led to humans was unique in origin (separate tree), yet could have easily led to a non-human form if chance had led it down another path. However, 'Abdu'l-Baha says that the early primitive forms that became human had signs of human potential that other animals lacked. Here is one such example:

> [W]e may acknowledge the fact that at one time man was an inmate of the sea, at another period an invertebrate, then a vertebrate and finally a human being standing erect. Though we admit these changes, we cannot say man is an animal. In each one of these stages are signs and evidences of his human existence and destination.[325]

This is an important distinction. If literally correct it means that there is a biological capacity in small primitive forms, and the capacity must be represented by certain genetic or structural characteristics. It must be observable. The distinction in mankind is spiritual in nature, but the spirit is attracted by physical composition. 'Abdu'l-Baha said,

> Let us give another, more subtle proof [that man has not come from the animal kingdom]... the completeness of each and every thing – that is, the completeness which you now see in man, or in other beings, with regard to their parts, members, and powers – arises from their component elements, their quantities and measures, the manner of their combination, and their mutual action, interaction, and influence. When all these are brought together, then man comes into existence.
>
> As the completeness of man stems entirely from the component elements, their measure, their manner of combination, and the mutual action and interaction of other beings – and since man was produced ten or a hundred thousand years ago from the same earthly elements, with the same measures and quantities, the same manner of composition and combination, and the same interactions with other beings – it follows that man was exactly the same then as exists now. This is a self-evident truth and cannot be doubted. And if a thousand million years hence, the component elements of man are brought together, measured out in the same proportion, combined in the same manner, and subjected to the same interaction with other beings, exactly the same man will come into existence.[326]

And in another place,

> the members, constituent parts, and composition that are found within man attract and act as a magnet for the spirit: The spirit is bound to appear in it… when the elements are composed and combined according to the noblest order, arrangement, and manner, the human spirit will appear and manifest itself therein.[327]

Christian Schwabe proposed the "genomic potential hypothesis" (see Ch. 4: Independent birth and genomic potential), which included a prediction that genomic potential of more complex life would be found in primitive forms. After his prediction, many such examples have been found (see Ch. 9: Genetic predisposition).

Combining genomic potential with the repeatability demonstrated in evolution, we can come to some interesting conclusions. There may be living primitive creatures that are so constrained in their potential that they are genetically pre-disposed to a certain mature form.

The same process that generated humans would have continued acting, and there may be existing primitive humans still making a living in nature today. In the next 20 years thousands of new genomes will be sequenced all over the world. We may find many more examples of genomic potential, and we may come across some uniquely human DNA.

If future discoveries identify an undeveloped human form crawling or swimming around, how should we regard it? Running this thought experiment there are two possibilities: it will never evolve to maturity because *Homo sapiens* beat it to the finish line, or it went down a path that didn't allow for its full potential. For example, having advanced speech and two free hands seems necessary for human civilization. Maybe elephants or dolphins have a similar capacity for the expression of virtues, but don't have the right feature set to carry forward an ever-advancing civilization. After all, elephant brains are similar in many ways to those of humans, but with more potential cognitive processing. They also have societies, bury their dead, and make war against enemies.

Humanity

What are the biological characteristics that made primitive humans distinctly human?

Anthropologists have long identified brain growth as the final step in human evolution. The use of tools, the freeing of the hands, the changing of the jaw, all led to brain development that distinguishes modern humans. Added recently to that list was having a long gestation and childhood. While some animals have a longer pregnancy than humans, those tend to come out walking and ready to take care of themselves in a short time. Humans, however, have a 40-week gestation followed by about 10 years of total reliance on parents for survival (or 30 years, depending on the child). From an evolutionary point of view, this comes with a great cost: investment from parents and greater risk of dying before reproduction. The evolutionary payoff from the gestation must be equally great, and it can be summed up in the one word that frequently passes the lips of zombies: *braaaains.*

A human-style brain requires a lot of blood and soaks up nearly a quarter of the energy in the body. Having warm blood also simplifies body functioning and fights off fungal infections. Thus, building an advanced brain seems to require warm blood, long placental gestation, long childhood, free hands, use of tools, family relationships, large tribes, and potentially many more constraints. The path to human intelligence seems to be very narrow.

This gives the impression that a series of random occurrences resulted in modern humans, and those same random occurrences could have happened to any creature, or never at all. The final step, according to modern theory, was that bipedalism freed the hands to use tools, which allowed for a changing diet, which reduced the huge muscles encompassing the head, which allowed the brain to fully develop.

However, even this string of events gives the impression of a latent capacity, because other species have experienced the same opportunities without the result of human intelligence. Something was different in the case of humans; an existing capacity was unleashed under the right conditions.

The traditional approach of contingency and variable outcomes ignores some of the integrated relationships between these occurrences. Tool use encouraged brain growth, which encouraged more and better tool use. Language allowed for social groups, which allowed for more complex language. Larger social groups allowed for longer childhood, which allowed for brain development, which allowed for better everything. As the brain developed, all those other processes benefited from it. David Bjorklund, professor of psychology, wrote about the interconnected nature of social complexity, big brains, and slow development in *Why Youth is Not Wasted on the Young*:

> There is no simple cause and effect here; the relation among these three factors is synergistic, with changes in one factor being both a cause and a consequence of changes in related factors. But social complexity was a required ingredient in human cognitive evolution. It exerted selection pressure for a bigger brain and a prolonged childhood, which in turn permitted increased levels of social complexity to be attained.[328]

Anthropologists were on the right track to target the brain for unique human traits. 'Abdu'l-Baha identified it as a pivotal organ for humans.

> As outer circumstances are communicated to the soul by the eyes, ears, and brain of a man, so does the soul communicate its desires and purposes through the brain to the hands and tongue of the physical body, thereby expressing itself.[329]

> The mind, therefore, has no place, although it is connected with the brain.[330]

> The human spirit is existent in the sight (eyes); it is also existent in the brain, which is the location of great functions and powers; it is also existent in the heart, which organ is largely connected with the brain or the center of the mind...[331]

He also reminds us that when it comes to brains, it's quality over quantity:

> The mere size of the brain has been proved to be no measure of superiority.[332]

So, the genetic and morphological characteristics that allowed for human consciousness, if present in primitive form, would likely be associated with the brain.

Human genome

With the completion of the human genome project in 2003, the genome was scoured not only for similarities, but also for differences between humans and other mammals. As it turns out, there are several uniquely human genes that play a role in brain development. These are often described as the result of "extremely rapid evolution" to explain their uniqueness.

- The lack of NDE1 causes microcephaly, in which the brain develops to only 10 percent of normal size.[333]
- Minor mutations of LAMC3 cause a deformity related to the folding of the brain. The folding is required to expand the surface area and allow for complex thought in a small space.
- A mutation of FOXP2 causes a severe speech disorder.
- A series of genes (e.g. CLOCK) are found in primates, but are regulated and expressed in a much more complex way in humans, supporting the brain's plasticity and ability to learn.[334]
- miR-941 is active in the brain and "controls our decision making and language abilities".[335]
- The human version of HARE5, a regulator of gene activity, leads to a larger brain and more neuron production.[336]
- HAR1, the most dramatic of the uniquely human regions, is expressed in neurons of the developing neocortex and "is of fundamental importance in specifying the six-layer structure of the human cortex".[337] Researchers at the University of Oxford said, "What is really interesting is that this is a special type of gene. It seems likely that it changes the way the brain is wired in some way."[338] Though a similar form of the sequence can be found in all mammals, the common version generally has no variation whatsoever

across any species, but in humans the sequence appears to have mutated greatly, and more rapidly than normally would be expected over a few million years of evolution.

It cannot be the mere presence of a gene sequence that makes a human. There are numerous complexities involved, but as 'Abdu'l-Baha described it is a particular physical composition of elements that attracts the spirit. There may be certain genetic sequences that will produce human intelligence given the right measure, balance, combination, and environmental influence. These sequences of uniquely human potential may indeed be found in primitive forms, such as worms, or found in other intelligent creatures, such as elephants. Such a discovery would provide an astonishing validation of repeated independent descent.

Review

Before the discovery of DNA, the evidence for evolution and common ancestry was presented with a few simple lines of observation and logic: the geological record shows an evolution from simple forms to more complex forms; breeding shows an example of the selection that takes place in nature; humans appear in the fossil record after many other animals were already established, indicating that humans are a branched lineage from animals; species improve and become hardier; and humans and other creatures have vestigial organs, some of which are very similar.

'Abdu'l-Baha seems to have disagreed with the conclusion that humans came from a branch of the ape family, and he gave three arguments for why the evidence does not support the conclusion (see Chapter 1).

First, he argued that animals appearing in the fossil record before humans is not proof that the humans came from animals. He said, "it is possible that man simply came into existence after the animal." And, "some [fruits] appear earlier in the season and others later. This priority is not a proof that the later fruit of one tree was produced from the earlier fruit of another."[339]

The biological basis for this has been established in Chapter 4. There is a mechanism for the growth of a new species from a seed-like form. Species forming in this way do not share common ancestry in any way as understood in the time of 'Abdu'l-Baha. Major phyla are probably not related, and the unique origins probably extend into the vertebrates. Some have speculated that unique multicellular formations are common and ongoing, extending to the level of family or genus. If confirmed with further investigation, then the later appearance in the fossil record would not always be an indication that a later species branched from an earlier species.

Second, he argued that the commonalities in vestigial organs is not proof that humans came from an animal. He said, "these minor traces and vestigial limbs might have some great underlying wisdom which the human mind has so far been unable to fathom." And, "the blackness of the pupil of the eye is due to its absorbing the rays of the sun, for if it were of another colour – say, uniformly white – it would not absorb these rays."

The biological basis for this has been established in Chapters 5 and 6. The evidence for common ancestry is based on the assumption of common useless remnants. A trait that does not go through a selection process would not converge in two unrelated organisms, just like two independent languages would not converge on mutual intelligibility. But, as 'Abdu'l-Baha suggests, when function and benefit is ascribed to a trait, it is no longer evidence for common ancestry. Added fitness and a selection process will converge on near-identical features and forms, or at least among a set of limited options. The implications of convergence and constrained options are now widely acknowledged and documented.

Third, he argued that humans changing and evolving from a primitive form is not proof that humans came from an animal. He said, "let us suppose that man once walked on all fours or had a tail: This change and transformation is similar to that of the fetus in the womb of the mother. Even though the fetus develops and evolves in every possible way before it reaches its full development, from the beginning it belongs to a distinct species."

The biological basis of this has been established in Chapters 8 and 9. Fossil evidence shows a separate line of human ancestry back to a more primitive form at least 7 million years ago. Fossils have recently confounded assumptions of ancestry and caused a rethink of proposed relatedness.

The discovery of DNA – decades after 'Abdu'l-Baha spoke – also created a biological basis for the potential of the embryo. When the embryo is a single cell it carries the genomic potential for a distinct species. In the same way, a new multicellular formation begins evolving within a limited set of developmental motifs and the broad outline of its evolutionary path is set. There is also evidence that new multicellular formations come with genomes that bias the evolutionary trajectory that the species will take. If independent descent is correct, the paradigm will shift over the next 20+ years as genomic potential is found in primitive forms.

After these arguments, 'Abdu'l-Baha summarizes,

> Just as man progresses, evolves, and is transformed from one form and appearance to another in the womb of the mother, while remaining from the beginning a human embryo, so too has man remained a distinct essence—that is, the human species—from the beginning of his formation in the matrix of the world, and has passed gradually from form to form. It follows that this change of appearance, this evolution of organs,

and this growth and development do not preclude the
originality of the species. Now, even accepting the
reality of evolution and progress, nevertheless, from the
moment of his appearance man has possessed perfect
composition, and has had the capacity and potential
to acquire both material and spiritual perfections
and to become the embodiment of the verse, "Let
Us make man in Our image, after Our likeness."[340]

It is entirely reasonable to focus on the social and spiritual
implications of 'Abdu'l-Baha's comments on human origin. If a
statement can be interpreted in two different ways, one of which
agrees with science and the other not, of course we should defer
to the one in agreement.

In the case of human origin, it is not quite as simple as two
ways to interpret the same thing. The references have a face value
that conflicts with universal common ancestry; the references are
extensive and direct, not passing comments; the language seems
non-symbolic.

While it is possible to conclude that 'Abdu'l-Baha was speaking
of spiritual potential and not material ancestry, the more apparent interpretation is of separate ancestry. I feel that it is no longer
necessary to conclude that the concept of independent or 'parallel'
descent is incompatible with science. In fact, the trend of discovery
has clearly been in the direction of agreement, and there are logical lines of inquiry that could entirely validate it in the next few
decades. This new understanding appears to me to have only been
possible since about 2015.

Final thought

In the fossil record humans appeared after other animals. In 'Abdu'l-Baha's time this was considered part of a body of evidence supporting common ancestry, because humans were presented as a recent divergent branch from the common tree. If humans existed potentially, then why did they take so long to appear in the mature form tens of millions of years after great apes? If species repeatedly form, why do we no longer see trilobites and dinosaurs?

The long evolutionary development of complex life can be compared to the development of a forest. When a fire or other catastrophe razes a very mature forest, it takes hundreds of years to fully restore it. The first thing to happen is a great deal of bacteria decomposing and processing raw elements. Next a series of other microscopic organisms come in to feed on the bacteria or provide some energy processing in the soil; these include protozoa, fungi, and nematodes. Then microscopic arthropods come on the scene to eat and recycle nutrients that are then used by the first plants to appear, which are opportunistic pioneer weeds that can grow quickly in poor soil, have a short life cycle, and produce huge numbers of seeds that disperse far and wide. Soon visible worms and arthropods appear. As biomass and diversity increase, the balance of fungi to bacteria levels out, so does the balance of nitrate and oxygen, then the weeds are out-competed by early grasses and forbs, ferns, shrubs, and more complex flowering trees. Biomass increases, soil diversity increases, the fungi/bacteria ratio increases, and vines appear amongst some deciduous saplings.[341]

The large deciduous trees grow slowly, penetrating far deeper into the soil than any previous plant. They soak up much of the available water and create a canopy that blocks out sunlight from any of their shorter counterparts. These trees also provide habitat for birds, reptiles, and small mammals that feed on fruits, vegetables, worms, and bugs. The fungi/bacteria ratio continues to grow as coniferous trees start to overtake the deciduous ones. The weeds have long been deprived of opportunities for growth. The conifers at first grow densely, but as they increase in size some naturally overtake their kindred. Their roots begin to penetrate tens of meters into the ground as they grow just as tall and several meters wide. In their mature state, they can live over one thousand years.

As the forest reaches maturity, its biology reaches an equilibrium that is favorable to larger animals. The small herbivores become the prey of larger predators, and as diversity and complexity increase, reproduction levels balance out to sustain healthy populations.

In this example, the 'seeds' of each plant were already in existence and ready to flourish when the environmental conditions were propitious. The elements of the mature forest started developing early on, but their growth was much slower than the rest. At first the coniferous shoots resembled weeds, and later appeared similar in height and appearance to the shrubs and fruit trees, but their genetic make-up made them destined to outstrip the rest, given enough time.

Now consider evolution over hundreds of millions of years. The ocean, then land, was first colonized by bacteria, then fungi, weeds, grasses, insects, and higher plants before animals appeared. It wasn't until the conditions for each stage existed that the next organism appeared. It was not possible for birds to develop until their habitat and lower food chain was present. The conditions for humans to appear in their current form took tens of millions of years longer than other placental mammals. At first humans may have resembled a worm or some other simple creature, and later appeared similar in height and appearance to primates, but their genetic make-up made them destined to outstrip the rest, given enough time.

Current consensus is that each stage came into existence through divergence from a previous stage, because they appeared in the fossil record at a later date. In the example of a maturing forest, each type of organism comes from an independent genealogical trace, and its mature appearance in the ecosystem is dependent on propitious conditions. I believe this is the case for biological evolution, but the 'seeds' represent a continuous stream of new creatures making the phase change towards multicellular life, each evolving along a path that becomes intolerant to significant change as they grow more complex. Within each tree, there may be many species from divergence among similar forms, but their differences are embellishments of the essence of the species.

LIST OF AUTHORS

From 1990 to 2009 these authors tried to address the Baha'i approach to evolution based on 'Abdu'l-Baha's comments.

- Craig Loehle: 'On human origins: A Baha'i perspective', in *Journal of Baha'i Studies*, vol. 2 (1990), no. 4;
- B. Hoff Conow: *The Baha'i Teachings* (Oxford: George Ronald, 1990);
- Charles Lerche: *Emergence* (London: Baha'i Publishing Trust, 1991);
- Gary Matthews: *The Challenge of Baha'u'llah* (Oxford: George Ronald, 1993);
- Craig Loehle: *On the Shoulders of Giants* (Oxford: George Ronald, 1994);
- William S. Hatcher: 'A Scientific Proof of the Existence of God' in *Journal of Baha'i Studies*, Vol. 5 (1994), no. 4;
- William S. Hatcher and John Hatcher: *The Law of Love Enshrined* (Oxford: George Ronald, 1995);
- Keven Brown: 'Are 'Abdu'l-Baha's views on evolution original?' in *Baha'i Studies Review*, vol. 7 (1997);

- Eberhard von Kitzing: 'Is the Baha'i view of evolution compatible with modern science?' in *Baha'i Studies Review*, vol. 7 (1997);
- Steven Friberg: 'Commentary' in *Baha'i Studies Review*, vol. 8 (1998);
- Mark A. Foster, 'Suggestions for Baha'i Hermeneutics', at *Baha'i Library Online*, (1999);
- Keven Brown and Eberhard von Kitzing: Evolution and Baha'i Belief: 'Abdu'l-Baha's Response to Nineteenth-Century Darwinism (Los Angeles: Kalimat Press, 2001);
- Fariborz A. Davoodi: 'Human evolution: Directed?', in *Baha'i Library Online* (2001);
- Courosh Mehanian and Stephan Friberg: 'Religion and evolution reconciled: 'Abdu'l-Baha's comments on evolution', in *Journal of Baha'i Studies*, vol. 13 (2003);
- Bahman Nadimi, 'Do the Baha'i Writings on evolution allow for mutation of species within kingdoms but not across kingdoms?', at *Baha'i Library Online*, (2004);
- Steven Phelps: 'Perspective: Crossing the divide between science and religion: A view on evolution', in *One Country*, vol. 19, no. 3 (June 2008);
- Ramin Neshati, 'Man is man: 'Abdu'l-Baha on human evolution', presentation at the Irfan Colloquia Session no. 74, Bosch Baha'i School, Santa Cruz, California, 24–27 May 2007, in *Lights of Irfan*, vol.10 (Wilmette, IL: Irfan Colloquia, 2009), pp. 295–310;
- Ian Kluge, 'Some Answered Questions: A Philosophical Perspective', in *Lights of Irfan*, vol. 10 (Wilmette, IL: Irfan Colloquia, 2009), pp. 149-274.
- Salman Oskooi, *When Science and Religion Merge: A Modern Case Study*, a thesis presented to the faculty of San Diego State University (2009).

APPENDIX B

TRANSLATIONS OF SOME ANSWERED QUESTIONS

The following table compares selections from *Some Answered Questions* in the original and revised translations. For a comparison of the full book, see

http://bahai-library.com/writings/abdulbaha/saq/diffs/abdulbaha_saq_comparison_table.html

Ch.par	1908 Translation	2014 Translation
46.1	We have now come to the question of the modification of species and of organic development – that is to say, to the point of inquiring whether man's descent is from the animal.	We now come to the question of the transformation of species and the evolutionary development of organs, that is, whether man has come from the animal kingdom.
46:2	This theory has found credence in the minds of some European philosophers, and it is now very difficult to make its falseness understood... For, verily, it is an evident error.	This idea has entrenched itself in the minds of certain European philosophers, and it is very difficult now to make its falsity understood... For in reality it is an evident error.

Ch.par	1908 Translation	2014 Translation
47:10	And in the same way, man's existence on this earth, from the beginning until it reaches this state, form and condition, necessarily lasts a long time, and goes through many degrees until it reaches this condition.	In like manner, from the beginning of man's existence on this planet until he assumed his present shape, form, and condition, a long time must have elapsed, and he must have traversed many stages before reaching his present condition.
47:10	But from the beginning of man's existence he is a distinct species. In the same way, the embryo of man in the womb of the mother was at first in a strange form;	But from the beginning of his existence man has been a distinct species. This is similar to the embryo of man in the womb of the mother: It possesses at first a strange appearance;
47:10	But even when in the womb of the mother and in this strange form, entirely different from his present form and figure, he is the embryo of the superior species, and not the animal; his species and essence undergo no change.	But even when it possesses, in the womb of the mother, a strange form entirely different from its present shape and appearance, it is the embryo of a distinct species and not of an animal: The essence of the species and the innate reality undergo no transformation at all.
47:11	Man was always a distinct species, a man, not an animal.	But man has always been a distinct species; he has been man, not an animal.

Ch.par	1908 Translation	2014 Translation
48:3	This is what the philosophers of the present state; this is their saying, this is their supposition, and thus their imagination decrees. So with powerful arguments and proofs they make the descent of man go back to the animal, and say that there was once a time when man was an animal, that then the species changed and progressed little by little until it reached the present status of man.	Such are the claims of the presentday philosophers. Such are their words, such are their claims, and such are the dictates of their imaginations. And so, after extensive research and armed with powerful arguments, they place man in the lineage of the animal, saying that at one time man was an animal, and that the species gradually changed and evolved until it reached the human degree.
48:4	But the theologians say: No, this is not so. Though man has powers and outer senses in common with the animal, yet an extraordinary power exists in him of which the animal is bereft.	But the divine philosophers say: No, this is not so. Although man shares the same outward powers and senses in common with the animal, there exists in him an extraordinary power of which the animal is deprived.
49:2	Briefly, this question will be decided by determining whether species are original or not – that is to say, has the species of man been established from its origin, or was it afterward derived from the animal?	Briefly, this question comes down to the originality or non-originality of the species, that is, whether the essence of the human species was fixed from the very origin or whether it subsequently came from the animals.

Ch.par	1908 Translation	2014 Translation
49:4	… it is possible that man came into existence after the animal… This priority does not prove that the later fruit of one tree was produced from the earlier fruit of another tree.	… it is possible that man simply came into existence after the animal… This priority is not a proof that the later fruit of one tree was produced from the earlier fruit of another.
49:7	We will state it more clearly. Let us suppose that there was a time when man walked on his hands and feet, or had a tail; this change and alteration is like that of the fetus in the womb of the mother. Although it changes in all ways, and grows and develops until it reaches the perfect form, from the beginning it is a special species.	To be more explicit, let us suppose that man once walked on all fours or had a tail: This change and transformation is similar to that of the fetus in the womb of the mother. Even though the fetus develops and evolves in every possible way before it reaches its full development, from the beginning it belongs to a distinct species.
49:8	To recapitulate: as man in the womb of the mother passes from form to form, from shape to shape, changes and develops, and is still the human species from the beginning of the embryonic period – in the same way man, from the beginning of his existence in the matrix of the world, is also a distinct species – that is, man – and has gradually evolved from one form to another.	To summarize: Just as man progresses, evolves, and is transformed from one form and appearance to another in the womb of the mother, while remaining from the beginning a human embryo, so too has man remained a distinct essence – that is, the human species – from the beginning of his formation in the matrix of the world, and has passed gradually from form to form.

Ch.par	1908 Translation	2014 Translation
49:8	Man from the beginning was in this perfect form and composition, and possessed capacity and aptitude for acquiring material and spiritual perfections…	… from the moment of his appearance man has possessed perfect composition, and has had the capacity and potential to acquire both material and spiritual perfections…
50:4	Therefore, it cannot be said there was a time when man was not. All that we can say is that this terrestrial globe at one time did not exist, and at its beginning man did not appear upon it.	We cannot say, then, that there was a time when man was not. At most we can say that there was a time when this earth did not exist, and that at the beginning man was not present upon it.
50:5	…it cannot be imagined that the worlds of existence, whether the stars or this earth, were once inhabited by the donkey, cow, mouse and cat, and that they were without man!	…it cannot be imagined that the world of existence, whether in the realms above or below, was once populated by cows and donkeys, cats, and mice, and yet was deprived of the presence of man.

Ch.par	1908 Translation	2014 Translation
51:2	The beginning of the existence of man on the terrestrial globe resembles his formation in the womb of the mother. The embryo in the womb of the mother gradually grows and develops until birth, after which it continues to grow and develop until it reaches the age of discretion and maturity. Though in infancy the signs of the mind and spirit appear in man, they do not reach the degree of perfection; they are imperfect. Only when man attains maturity do the mind and spirit appear and become evident in utmost perfection.	The beginning of the formation of man on the terrestrial globe is like the formation of the human embryo in the womb of the mother. The embryo gradually grows and develops until it is born, and thereafter it continues to grow and develop until it reaches the stage of maturity. Although in infancy the signs of the mind and the spirit are already present in man, they do not appear in a state of perfection, and remain incomplete. But when man attains maturity, the mind and the spirit manifest themselves in the utmost perfection.

Ch.par	1908 Translation	2014 Translation
51:3	So also the formation of man in the matrix of the world was in the beginning like the embryo; then gradually he made progress in perfectness, and grew and developed until he reached the state of maturity, when the mind and spirit became visible in the greatest power. In the beginning of his formation the mind and spirit also existed, but they were hidden; later they were manifested.	Likewise, at the beginning of his formation in the matrix of the world, man was like an embryo. He then gradually progressed by degrees, and grew and developed until he reached the stage of maturity, when the mind and the spirit manifested themselves in the utmost perfection. From the beginning of his formation, the mind and spirit existed, but they were hidden and appeared only later.
51:5	Similarly, the terrestrial globe from the beginning was created with all its elements, substances, minerals, atoms and organisms; but these only appeared by degrees: first the mineral, then the plant, afterward the animal, and finally man. But from the first of these kinds and species existed, but were undeveloped in the terrestrial globe, and then appeared only gradually.	Similarly, the terrestrial globe was created, from the beginning, with all its elements, substances, minerals, parts, and components, but these appeared only gradually: first the mineral, then the plants, then the animals, and finally man. But from the beginning, these kinds and species were latent in the earthly realm and appeared gradually thereafter.

PROMULGATION OF UNIVERSAL PEACE

Revised translation

Below is part of a talk by 'Abdu'l-Baha at the Open Forum in San Francisco, 10 October 1912. This revised provisional translation is by Keven Brown, parts of which were published in Evolution and Baha'i Belief (2001), pp. 98-99 and pp. 235-236.

Persian Source: Majmú'iyyih Khitabat-i 'Abdu'l-Baha (Collection of the Talks of 'Abdu'l-Baha), (Hofheim-Langenhain: Baha'i Verlag, 1984) 2:301-304. Reprint of the three original volumes printed in Egypt 99 BE (1942/43) and 1340 AH (1921) and Tehran 127 BE (1970/71).

English notes of the oral interpretation: Promulgation of Universal Peace (1982), pp. 358-359; or (2012), pp. 505-507.

Briefly, the evidences of the intellect of man are manifest and clear. Man is man by reason of this intellectual faculty. Therefore, the animal kingdom is other than the human kingdom. Notwithstanding this, the philosophers of the West have adduced evidences to demonstrate that man had his origin in the animal kingdom. They say that originally he was one of the creatures swimming in the sea; afterwards he emerged from the water to live upon the land, and he became a vertebrate. Then appendages appeared. At first he walked on four legs, then he became a two-footed animal [walking erect], and that two-footed animal is man. In other words, he was transferred from one state to another until he reached this human shape and form. They say that the manner of man's formation can be compared to the links of a chain, which are connected one to another. However, between man and the ape one link is missing. Great scientists and philosophers have searched for it, some even devoting their whole lives to solving this problem, but until now they have been unable to find that missing link.

But their greatest proof is this: vestiges of members are present in man which also belong to certain animals. From this they deduce that during the course of centuries and ages man has evolved so that now those members have become lost. For example, the snake has one vestige that indicates at one time it had appendages. But when it started living in holes underground it had no need for such appendages; finally, little by little, those members vanished but vestiges of them remain today. This proves that at one time it possessed appendages.

Likewise, man possesses vestiges which originally had another shape, and now their shape has changed. For example, there is a vestige in the body of man at the end of his spinal column that indicates at one time man had a tail. After he began to walk erect, gradually the tail vanished. By this theory Western philosophy has finally culminated in the tail of the ape and has become perplexed and bewildered in pursuit of the missing link.

The philosophers of the East say: If the human body was originally not in its present composition, but was gradually transferred from one state to another until it appeared in its present form [as the philosophers of the West say], then we would postulate that although at one time it was a swimmer and later a crawler, still it was human, and its species has remained unchanged. The proof for this is that the human embryo is at first a mere germ. Gradually the hands and feet appear and the lower limbs become separated from each other, and it is transferred from one form to another, one shape to another, until it becomes born with this shape and appearance. But from the time it was in the womb in the form of a germ, it was the species of man and not the embryo of other animals. It was in the form of a germ, but it progressed from that form to this most beautiful form. Therefore, it is clear that the species is preserved.

Provided that we assent [to this theory] that man was at one time a creature swimming in the sea and later became a four-legged, assuming this to be true, we still cannot say that man was an animal. Proof of this lies in the fact that in the stage of the embryo man resembles a worm. The embryo progresses from one form to another, until the human form appears. But even in the stage of the embryo he is still man and his species remains unchanged.

The link which they say is lost is itself a proof that man was never an animal. How is it possible to have all the links present and that important link absent? Though one spend this precious life searching for this link, it is certain that it will never be found.

APPENDIX D

AN UNAUTHENTICATED REFERENCE

I chose to omit an unauthenticated quotation from 'Abdu'l-Baha that seemed out of place with the rest of the references to evolution. It is printed in Anjam Khursheed, *Science and Religion: Towards the Restoration of Ancient Harmony* (London: Oneworld, 1987), p. 90.

Moses taught that the world was brought into existence in the six days of creation. This is an allegory, a symbolic form of the ancient truth that the world evolved gradually. Darwin can refer to Moses for his theory of evolution. God did not allow the world to come into existence all at once, rather the divine breath of life manifested itself in the commanding Word of God, *Logos*, which engendered and begot the world. We thus have a progressive process of creation, and not a one-time happening. Moses' days of creation represent time spans of millions of years. From Pythagoras to ibn-i-Sina (known as Avicenna) to the "faithful brothers from Basra", through Darwin and to the blessed manifestations of the Bab and Baha'u'llah, both scholars and Prophets have testified to the progressive creative action of the *Logos* (divine breath of life). The Darwinian and monistic theories of evolution and the origin of species are not materialistic, atheistic ideas, they are religious truths which the godless and the deluded have unjustifiably used in their campaign against religion and the Bible.

Khursheed's reference says,

Conversation between 'Abdu'l-Baha and Dr Fallscheer, recorded in *Sonne der Wahrheit*, No. 1, March 1921, p. 9.

I questioned the Research Department at the Baha'i World Centre and received the following reply on 16 June 2011.

... the statement in question is part of an English translation of a talk of 'Abdu'l-Baha given in Arabic in the summer of 1910 and published in German in *Sonne der Wahrheit*. The transmitter, identified as "Dr. F.", was Dr. Josephine Fallscheer, a doctor who attended to members of the household of 'Abdu'l-Baha and the author of other pilgrim notes dated to this period which were also published in *Sonne der Wahrheit*. As the original Arabic text has not been located, the statement cannot be authenticated at this time.

Although I did use a short part of the quotation ("Moses' days of creation represent time spans of millions of years"), I felt that leaving the complete reference in the appendix would be the appropriate place for interested readers.

APPENDIX E
REFERENCES TO EVOLUTION

List prepared by the Baha'i World Centre, 2005

The following list was prepared in July 2005 by the Research Department at the Baha'i World Centre and has been provided in response to several queries on the Baha'i perspective on evolution. It was attached to a letter to an individual dated 4 September 2005.

Some Answered Questions (Wilmette: Baha'i Publishing Trust, 1984) See chapters 46, 47, 48, 49, 50, 51, and 64.

The Promulgation of Universal Peace: Talks Delivered by 'Abdu'l-Baha during His Visit to the United States and Canada in 1912 (Wilmette: Baha'i Publishing Trust, 1982)

See, for example, pages 67–69; 225–26; 355–61. Many useful references are to be found under such index terms as "Creation", "Evolution", and "Existence".

Paris Talks: Addresses given by 'Abdu'l-Baha in Paris in 1911–1912 (London: Baha'i Publishing Trust, 1979)

See, for example, pages 88–94; pages 96–99.

Abizadeh, Arash, "Commentary: 'On Human Origins: A Baha'i Perspective'", published in *The Journal of Baha'i Studies*. volume 3, number 1, pages 67–73.

Brown, Keven, "Response to Commentary on 'On Human Origins'", published in *The Journal of Baha'i Studies*, volume 5, number 4, pages 59–62.

Loehle, Craig, "On Human Origins: A Baha'i Perspective", published in *The Journal of Baha'i Studies*, volume 2, number 4, pages 45–58.

Loehle, Craig, *On the Shoulders of Giants* (Oxford: George Ronald, 1994) Chapter 4 deals with evolution in the Baha'i perspective.

Mehanian, Courosh and Friberg, Stephen R., "Religion and Evolution Reconciled: 'Abdu'l-Baha's Comments on Evolution", published in *The Journal of Baha'i Studies*, volume 13, number 1/4, pages 55–93.

APPENDIX F

SHOGHI EFFENDI

Extracts from Letters Written on Behalf of Shoghi Effendi on the Baha'i Perspective on Evolution*

The following list was prepared by the Research Department at the Baha'i World Centre and has been provided in response to several queries on the Baha'i perspective on evolution. It was attached to a letter to an individual dated 4 September 2005. The footnotes are from the original memorandum.

With reference to the statement to the effect that man has lived as man upon the earth only 300,000 years: I am directed by the Guardian to inform you that this has never originated from him, and is a misquotation. Questions such as this are for scientists to investigate and decide upon.

(23 November 1937)

---

* Letters are addressed to individuals unless otherwise specified.

Man, as a distinct species, has been specifically and for the first time, asserted by 'Abdu'l-Baha in His "Some Answered Questions".[*]

(30 JULY 1941, TO A COMMITTEE OF A
NATIONAL SPIRITUAL ASSEMBLY)

We cannot prove man was always man for this is a fundamental doctrine, but it is based on the assertion that nothing can exceed its own potentialities, that everything, a stone, a tree, an animal and a human being existed in plan, potentially, from the very "beginning" of creation. We don't believe man has always had the form of man, but rather that from the outset he was going to evolve into the human form and species and not be a haphazard branch of the ape family.

You see our whole approach to each matter is based on the belief that God sends us divinely inspired Educators; what They tell us is fundamentally true; what science tells us today is true, tomorrow may be entirely changed to better explain a new set of facts.[†]

(7 JUNE 1946)

There is nothing unreasonable in the Master's statement you quote, page 220, "The Promulgation of Universal Peace": He expounds the idea that man was always potentially man, which is just another way of saying the Cause contains the power to produce the effect;

---

[*] This answer was given in response to a question concerning whether the Qur'án presents a concept of biological evolution and reveals man as a distinct species or whether this explanation was first revealed in the Bahá'í Era.

[†] This answer was written in response to a question about statements of 'Abdu'l-Bahá in *The Promulgation of Universal Peace*. The enquirer asked how the Master's statement that, "even when the embryo resembled a worm, it was human in potentiality, not animal", might be proved; and he commented on 'Abdu'l-Bahá's statement about "the missing link".

in this planned and integrated universe, he was part of the plan from the beginning, so to speak.*

(24 FEBRUARY 1947)

We Baha'is do not believe in Genesis literally. We know this world was not created in seven days, or six, or eight, but evolved gradually over a period of millions of years, as science has proved.

(29 OCTOBER 1949)

The Baha'i Faith teaches man was always potentially man, even when passing through lower stages of evolution.

(4 OCTOBER 1950)

---

* This statement appears on p. 225 in the 1982 edition of this book.

MEMORANDUM 2014

Memorandum from the Research Department regarding Shoghi Effendi on human evolution

The following was prepared by the Research Department at the Bahá'í World Centre in 2014. It was in response to an individual inquiring for a copy of the incoming communication to Shoghi Effendi that precipitated a response from him on 7 June 1946. The response by Shoghi Effendi includes the phrase "from the outset he was going to evolve into the human form and species and not be a haphazard branch of the ape family". The footnotes are from the original memorandum.

16 April 2014

We have located the incoming letter to Shoghi Effendi that Mr. ... requests. The letter, dated 16 May 1946, was written by J. W. Freudenberg, a friend of the Cause. In general, communications written to the Head of the Faith are considered to be confidential communications, and therefore a copy of the letter is not provided. However, we have prepared a summary of the relevant sections of the communication, which we present below...

Summary of Letter Dated 16 May 1946 from J. W. Freudenberg to Shoghi Effendi

In his letter, written from Auckland, New Zealand, Dr. J. W. Freudenberg takes issue with a number of statements attributed to 'Abdu'l-Baha. Quoting "But at all times, even when the embryo resembled a worm, it was human in potentiality and character, not animal",[*] he asks how this can be proved, stating that the theory of evolution is "recognised by all authorities" and has itself been proved in various ways by Darwin, Wallace and their successors. In reference to the statement, "in the protoplasm, man is man. Conservation of species demands it",[†] he insists that there is no such demand in nature at all, citing as evidence the fact that so many species have in the past become extinct. Dr. Freudenberg goes on to question statements regarding "the missing link"—namely, the fact that it has not yet been found indicates that "man has never been an animal", and moreover, that it "will never be found".[‡] He asserts that 'Abdu'l-Baha cannot know that the missing link will never be found, that anthropologists may have already found it, and that, in any event, the theory of evolution does not depend on its discovery as the theory

---

[*] *The Promulgation of Universal Peace: Talks Delivered by 'Abdu'l-Bahá during His Visit to the United States and Canada in 1912*, rev. ed. (Wilmette: Bahá'í Publishing, 2012), p. 507.

[†] Ibid.

[‡] Ibid.

has enough evidence to support it. Dr. Freudenberg then expresses disagreement with the statements that in the world of creation human and animal are to be classified separately[*] and that man is "the highest specialized organism of visible creation, embodying the qualities of the mineral, vegetable and animal plus an ideal endowment absolutely absent in the lower kingdoms".[†] Paraphrasing the Bahá'í view as, "the human gene and genius was already endowed in the first protoplasm", he contrasts it with the "Darwinistic" view, which holds that man, with his intellect, developed slowly but continually from "the first monera and amoeba" to the animal state of today. Moreover, he maintains that according to this latter perspective, the intellects of man and of animal differ only in degree, not in principle. He continues by taking issue with the statement that "man breaks the laws of nature",[‡] arguing that by flying and by riding in submarines (examples given in *The Promulgation of Universal Peace*[§]) man does not break the laws of nature but rather works within them, "because nature has permitted us to do so by means of our intellect which is also in the frame of nature and natural development". Finally, Dr. Freudenberg contests the statement that nature is "devoid of memory",[¶] stressing that nature "has an excellent memory, often over millions of years". He shares, as an example, the case of young storks which, having been separated from their parents, were able to find the migration routes of their parents.

---

[*] Ibid., pp. 39-40.

[†] Ibid., p. 40.

[‡] Ibid., p. 508.

[§] See ibid.

[¶] Ibid., p. 509.

There are two additional points regarding the letter that warrant mentioning. The first is that though in his letter Dr. Freudenberg does not indicate a source for the statements to which he refers, it is likely that it was *The Promulgation of Universal Peace*, which, as explained in the 2012 edition (Wilmette: Baha'i Publishing), was first published in two volumes in 1922 and 1925. The second is that the statements with which Dr. Freudenberg takes issue derive from two talks given by 'Abdu'l-Baha, for neither of which an original text has been found.

To further assist… in his attempt to better contextualize Shoghi Effendi's reply to Dr. Freudenberg, he may find it helpful to consider the following passage from a letter dated 15 September 2013 written on behalf of the Universal House of Justice to an individual believer:

> In academia, at the present time, the common ground of understanding in relation to all aspects of human behaviour is generally materialistic. In this connection, it is important to bear in mind the distinction between scientific findings in themselves and the materialistic philosophical interpretation commonly applied to them. While some individuals, aware of certain scientific perspectives pertaining to evolution or neuroscience, assert that there is no God and that the soul does not exist, such an assertion is not the same as saying that science has proven that there is no God and no soul. Just as you have found authors, whether scientists or philosophers, who have reached materialistic conclusions, there are many others who feel that such conclusions are unwarranted and that the current, dominant materialistic perspective cannot adequately account even for the existence of human consciousness. Thus, some important concepts set forth by 'Abdu'l-Baha, for example, that through its higher powers the human

mind is able to transcend nature and that the appearance of this rational capacity in the physical world was deliberate rather than arbitrary, are subjects of philosophical and theological inquiry that have not been, and perhaps can never be, resolved by science alone.

Also of interest is the following comment on the relationship between 'Abdu'l-Baha's explanations of the evolutionary process and current scientific theory. A letter dated 8 October 2012 written on behalf of the Universal House of Justice to an individual believer states:

Regarding your comments about the topic of evolution as presented in *Some Answered Questions*, individuals are, of course, free to come to their own understanding of the Sacred Writings. However, it is not necessary to conclude that 'Abdu'l-Baha was advocating a scientific position that contradicts contemporary evolutionary theory. Indeed, a number of Baha'is have looked at this subject from different perspectives. Some suggest that the statements in *Some Answered Questions* are not in conflict with modern scientific principles. Such an argument is presented, for example, in the article "Religion and Evolution Reconciled: 'Abdu'l-Baha's Comments on Evolution" by Courosh Mehanian and Stephen R. Friberg, published in *The Journal of Baha'i Studies*, volume 13, number 1/4, March–December 2003, pages 55–93, and accessible online at www.bahaistudies. ca/journal/files/jbs/13.14%20Mehanian%20&%20Friberg.pdf. Another viewpoint suggests that the Master was referring in His Writings to the philosophical and social implications of Darwin's scientific findings in light of the discourse current at that time rather than proposing an alternative scientific explanation. This approach is represented by Keven Brown's monograph, "Evolution and Baha'i Belief", published in *Evolution and Baha'i Belief: Studies in the Babi and Baha'i Religions,* volume 12, pages 5–133, which may be accessed at the following site: bahai-library.com/brown_forward_article_evolution.

LETTER 2016

Letter on Behalf of the Universal House of Justice Regarding Science and Sacred Scripture

21 February 2016

[To an individual]

Dear Baha'i Friend,

Your email letter 19 June 2015 including your thoughtful questions about a paragraph regarding evolution in the foreword to the 2014 edition of *Some Answered Questions* has been received by the Universal House of Justice, which has asked us to convey to you the following in reply. The delay in our response, which is due to the pressure of work at the Baha'i World Centre, is regretted.

As you have observed, the purpose of the paragraph in question, which the House of Justice approved for inclusion in the foreword, does not limit how a Baha'i, as an individual, may personally choose to interpret the Sacred Writings. Yet, the paragraph does not insist that science is "absolute truth", nor, as you seem to conclude, does it attempt to "apologize" for 'Abdu'l-Baha's statements. Rather, recognizing that He would not make a statement that contradicts reality, the paragraph encourages the friends to use all of the relevant texts on the subject as well as the most accurate and reliable picture of reality that science can provide to try to understand what 'Abdu'l-Baha actually is conveying.

It is evident that there are instances throughout history when statements made in the Sacred Scriptures that conflicted with the scientific views of the time were confirmed by science itself centuries later. There also may well be statements in the Writings about the material world the veracity of which will be proven by science in future. The notion of scientific "truth" does not encompass every claim or theory asserted in the name of science. But while a great deal of scientific discourse is tentative and subject to change, some scientific statements are accurate and reliable descriptions of reality, and those findings are not in conflict with true religion, that is, with the Revelation and its authorized interpretations. It is for this reason that 'Abdu'lBaha emphasizes that religious beliefs should be weighed in the light of science and reason, so that personal interpretations of the meaning of the Revelation, which are also fallible and subject to change, do not lead to incorrect conclusions.

The Master's statements on evolution are subtle and complex and must be understood within the context of the entirety of the Baha'i teachings, because His statements are both predicated upon and coherent with those teachings. In the passages found in *Some Answered Questions*, as well as in numerous other Tablets and talks, 'Abdu'l-Baha elaborates upon the principle of the harmony of science and religion, observes that human beings and animals have in common the same physical nature, emphasizes that it is the mind and the soul that distinguish humanity, and rejects the idea that human beings are merely animals, a haphazard accident, and captives of nature trapped in the struggle for existence. In light of all such statements, it is possible for a Baha'i to conclude that one can disagree with the materialistic philosophical interpretation of scientific findings – that man is merely an animal and a random expression of nature – without contesting the scientific findings themselves, such as those in genetics which are incompatible with a concept of "parallel" evolution.

Of course, different individuals, using their rational powers to reach personal interpretations of scientific findings and the meaning of Sacred Texts, may come to different conclusions on different questions. This is the inevitable outcome of the independent investigation of truth. On certain matters, there may for a time be a degree of ambiguity; on others, an exchange of views conducted in a consultative spirit may make the truth evident. Yet, in their efforts to explore the ocean of Baha'u'llah's Revelation, the House of Justice hopes that the friends will guard against two extremes. The first is to simply dismiss the truths found in the Revelation owing to a dogmatic attachment to materialistic interpretations of scientific findings. The second is to assume that in every instance where one's personal understanding of the teachings conflicts with scientific findings, it is these findings that must change in future, for such a posture would place Baha'is in the position of constantly contending with science. Both of these extremes are incompatible with the Baha'i principle of the harmony of science and religion.

As you consider this matter, you may find of interest the work of those believers who have attempted to correlate 'Abdu'l-Baha's statements with contemporary science, such as the article "Religion and Evolution Reconciled: 'Abdu'l-Baha's Comments on Evolution" by Courosh Mehanian and Stephen R. Friberg, published in *The Journal of Baha'i Studies*, volume 13, number 1/4, pages 55–93, which may be found at bahai-studies.ca/past-issues.

With loving Baha'i greetings,
Department of the Secretariat

LETTER 2019

Letter on behalf of the Universal House of Justice regarding publishing

Guidance was requested from the Universal House of Justice regarding the wisdom of publishing this book. Several questions were asked: does the Foreword to Some Answered Questions and the 2016 letter close all discussion of the subject of parallel evolution; is it possible to understand Some Answered Questions in a light different from that described in the Foreword; would it be possible to investigate these issues now; and what is the role of a publisher and review committee in such a situation? The following was the response.

17 February 2019

Your email letter of 3 January 2019 asking whether you should proceed with publication of the book titled *On the Originality of Species...* in light of concerns expressed to you that the book would be controversial and that it contradicts the foreword to the new translation of *Some Answered Questions*, has been received by the Universal House of Justice. We have been asked to respond as follows.

As mentioned in the 21 February 2016 letter written on behalf of the Universal House of Justice, the relevant paragraph in the foreword to *Some Answered Questions* "does not limit how a Baha'i, as an individual, may personally choose to interpret the Sacred Writings." Rather, it points out that in light of all of the statements of 'Abdu'l-Baha on the subject, it is possible for a Baha'i to disagree "with the materialistic philosophical interpretation of scientific findings—that man is merely an animal and a random expression of nature—without contesting the scientific findings themselves, such as those in genetics which are incompatible with a concept of "parallel" evolution." It further acknowledges: "On certain matters, there may for a time be a degree of ambiguity; on others, an exchange of views conducted in a consultative spirit may make the truth evident."

Provided that individuals do not, in their written works, misrepresent the Baha'i teachings, they have a right to express their opinions even if those opinions prove to be mistaken. Thus the friends are free to express their own personal views about the teachings in relation to a particular scientific theory or body of thought, but what they cannot assert is that these constitute the *Baha'i* view on the matter. If, however, the author has not made a clear distinction between his own opinions and the statements of 'Abdu'l-Baha, ascribing to 'Abdu'l-Baha ideas or arguments which represent, rather, his own understanding of what 'Abdu'l-Baha meant, that would be a cause for reviewers to question the accuracy of the representation of the teachings.

When it comes to publication, other factors become relevant. As you are aware, the purpose of review is to ensure that a work does not represent the teachings in an inaccurate or undignified manner, and that its publication would not be untimely and cause harm, but it is not within the mandate of review to offer assessments of a manuscript's literary quality or its broader value. And although reviewers may win the gratitude of authors by calling attention to such things as occasional grammatical or spelling errors—or other errors that do not concern the teachings—approval should not be refused on such grounds. Thus works that are unsuitable for publication for various reasons may pass review. Although reviewers are required to pay close attention to the accuracy of historical works, as these directly concern the Faith itself, to ask them to judge the validity of a manuscript from a scientific perspective would exceed their mandate. The decision to publish a work is a matter for the publisher's judgement. The publisher must ensure that a book on a scientific topic is well grounded in fact; it must decide whether it would be a valuable contribution to the relevant discourse, as well as whether it meets the house standards of quality. As the book will be published under its imprint, it must be prepared to accept the response the book may elicit once released, recognizing that if a work about science is perceived to be insufficiently grounded, it will be judged accordingly by readers and may adversely impact the publisher's reputation.

With loving Baha'i greetings,
Department of the Secretariat

GLOSSARY

Abiogenesis – the natural origin of living things from non-living material.

Base pair – a bonded pair of nucleotides that form the double helix of DNA. The complementary bonds in DNA are adenine-thymine and guanine-cytosine.

bya – billion years ago.

Codon – a sequence of three adjacent nucleotides that determine a specific amino acid in the genetic code.

DNA – Deoxyribonucleic acid, a nucleic acid that carries genetic information in a twisted double helix.

Epigenetics – the study of heritable traits that are caused by DNA methylation, rather than a change in base pairs.

ERV – Endogenous retrovirus, a virus that inserts its own DNA into a germline cell and is carried into the host's offspring.

Eukaryote (adj. eukaryotic) – an organism whose cells have a nucleus and other organelles inside a membrane.

Genotype – the genetic constitution of an individual organism.

Heterosis (adj. heterotic) – the improved biological quality in a hybrid offspring.

Hominid – a species in the family Hominidae, which includes orang-utans, gorillas, chimpanzees, humans, and their extinct relatives.

Hominin (formerly known as hominid) – a species of the tribe Hominini, which includes modern humans, their direct ancestors, and their extinct relatives.

Karyotype – the genetic structure that must closely match in mating populations, such as the quantity, size, and arrangement of chromosomes and loci.

Locus (pl. loci) – the position of a given gene or genetic marker on a chromosome.

Meiosis (adj. meiotic) – the process of cell division in all sexually reproducing organisms.

Microbiome – the complete genetic content of all the microorganisms inhabiting an organism in a symbiotic relationship.

Mitochondria – an organelle with genetic material that processes energy in most eukaryotic cells.

Mitosis (adj. mitotic) – the process of cell division by which the nucleus divides, resulting in a copy of the parent nucleus.

mya – million years ago.

Nucleus – an organelle that holds most of the genetic content in eukaryotic cells.

Nucleotide – the molecules that forms the building blocks of the genetic code. DNA uses adenine, cytosine, guanine, and thymine. RNA uses uracil to replace thymine.

Phenotype – observable characteristics of an organism.

Prokaryote – a microorganism whose cell lacks a membrane, including Archaea and Bacteria.

RNA – Ribonucleic acid, a nucleic acid involved in turning the genetic code into proteins.

Symbiogenesis – the formation of a new organism through the merging of freeliving organisms.

Telomere – sections of DNA at the ends of a chromosome.

SUGGESTED READING

- W. Ford Doolittle. 'Uprooting the tree of life', in *Scientific American*, vol. 282, no. 2 (February 2000), pp. 90–95.
- Simon Conway Morris. *Life's Solution: Inevitable Humans in a Lonely Universe*. Cambridge: Cambridge University Press, 2003.
- Frank Ryan. *Virolution*. London: Harper Collins, 2009.
- James Shapiro. *Evolution: A View from the 21st Century*. New Jersey: FT Press Science, 2011.
- Simon Conway Morris. *Runes of Evolution: How the Universe Became Self-Aware*. Templeton Press; 2015.
- Jonathan B. Losos. *Improbable Destinies: Fate, Chance, and the Future of Evolution*. New York: Riverhead Books, 2017.
- *ScienceDaily*. Science news aggregator that summarizes science publications into easily digestible articles. https://www.sciencedaily.com/

ENDNOTES

Chapter 1: Evolution and the Baha'i Faith

1   'Abdu'l-Bahá, *Some Answered Questions* (Haifa: Bahá'í World Centre, new translation 2014), ch. 83;

   The Promulgation of Universal Peace: Talks Delivered by 'Abdu'l-Baha During His Visit to the United States and Canada in 1912 (Comp. H. MacNutt. Wilmette, IL: Bahá'í Publishing Trust, Centenary Ed. 2012) pp. 27-30, 353-355.

2   'Abdu'l-Bahá, *The Promulgation of Universal Peace*, p. 323.

3   Ibid. p. 86.

4   Ibid. p. 555.

5   'Abdu'l-Bahá, *Paris Talks: Addresses Given by 'Abdu'l-Bahá in 1911*, (Wilmette, IL: Bahá'í Publishing Trust, Centenary Ed. 2011) no. 44.

6   Letter on behalf of Shoghi Effendi to an individual, 30 September 1950, in *Messages to Canada*, pp. 133-134.

7   'Abdu'l-Bahá, *Makátib-i 'Abdu'l-Bahá*, vol. 3 (Collected Letters) (Cairo 1921) pp. 172-173. Provisionally translated by Keven Brown and cited in Brown, *Evolution and Baha'i Belief* (Los Angeles: Kalimát Press, 2001), p. xvi. Full provisional translation available at https://bahai-library.com/abdul-baha_brown_science_religion

8   *ibid*

9   Pew Research Center, *Americans, Politics, and Science Issues* (1 July 2015), chapter 4.

10 Gary Matthews, The Challenge of Bahá'u'lláh (Wilmete, IL:

Bahá'í Publishing, 2005), p. 93. [First edition published 1993 by George Ronald]

11   Shoghi Effendi (1944). *God Passes By*, p. 242. Wilmette, IL: US Baha'i Publishing Trust.

12   Charles Darwin, *On the Origin of Species* (1859; London: John Murray, 6th ed. 1872), p. 425.

13   'Abdu'l-Bahá, *Some Answered Questions*, no. 47, p. 207.

14   'Abdu'l-Bahá, *The Promulgation of Universal Peace*, p. 653.

15   'Abdu'l-Bahá, quoted in Anjam Khursheed, *Science and Religion: Towards the Restoration of Ancient Harmony* (London: Oneworld, 1987), p. 90. The original Arabic text for this quotation has not been located, and the statement cannot be authenticated, according to a letter to the author from the Department of the Secretariat, Bahá'í World Centre. For more details see Appendix D: An Unauthenticated Reference.

16   'Abdu'l-Bahá, *Some Answered Questions*, no. 30, p. 138.

17   Letter on behalf of the Universal House of Justice to individual believers, 3 June 1982, in The Universal House of Justice, *Messages from the Universal House of Justice 1963–1986* (Comp. Geoffry W. Marks. Wilmette, IL: Bahá'í Publishing Trust, 1996), no. 330, pp. 545–6.

Available at: http://bahai-library.com/uhj_infallibility_abdulbaha.

18   Letter on behalf of Shoghi Effendi to an individual believer, 26 February 1933, in *Lights of Guidance: A Bahá'í Reference File* (Comp. Helen Hornby. New Delhi: Bahá'í Publishing Trust, 1994 edition), no. 1434.

19   Letter on behalf of Shoghi Effendi to the United States Publishing Committee, 29 December 1931, in *Lights of Guidance*, no. 1435.

20   Letter on behalf of the Universal House of Justice, 23 March 1987.

Available at: https://bahai-library.com/uhj_authenticity_some_texts

21   Shoghi Effendi commented on the English translations and authenticity: "The translations in *Promulgation of Universal Peace* are too inaccurate, in some places, to use them as an absolute basis for discussing some point, and he has not time at present to go over them, so the best thing is to put down any discrepancies as being due to this." (19 March 1946 to an individual, cited from a Memorandum

of the Research Department of the Universal House of Justice, dated 19 March 1995).

And in another letter: "Regarding the report of Promulgation of Universal Peace... Ultimately the Persian originals must be the basis for authentic statements made by the Master, but this will require time, scholars and research work not available at the present time." (5 July 1950 to a National Spiritual Assembly)

These were published in von Kitzing's *Origin of Complex Order in Biology* (1999 version), chapter 1, footnote 3.

22 "A verbatim record in Persian of His talks would of course be more reliable than one in English, because He was not always accurately interpreted." (On behalf of Shoghi Effendi, 24 October 1947 to the National Spiritual Assembly of the British Isles)

Published in "Unfolding Destiny: The Messages from the Guardian of the Bahá'í Faith to the Bahá'í Community of the British Isles" (London: Bahá'í Publishing Trust, 1981), p. 208.

23 Mahmud-i-Zarqani, *Mahmud's Diary* (Oxford: George Ronald, 1998), p. 62. See also Mahmud-i-Zarqani, *Kitab-i-Badayi'u'l-Athar* (1914), Vol. 1, pp. 54,58.

24 Letter on behalf of Shoghi Effendi to the United States Publishing Committee, 29 December 1931, in *Lights of Guidance*, no. 1435.

25 Letters on behalf of Shoghi Effendi to individual believers, 9 February 1932, 9 June 1932 and 9 November 1932, in *The Compilation of Compilations* (prepared by the Universal House of Justice 1963–1990. 2 vols. Sydney: Bahá'í Publications Australia, 1991), vol. I, nos. 460, 464 and 467.

26 'Abdu'l-Bahá, *Some Answered Questions*, no. 49.

27 'Abdu'l-Bahá, *The Promulgation of Universal Peace*, pp. 506-507.

28 Shoghi Effendi, *Arohanui: Letters from Shoghi Effendi to New Zealand* (Suva, Fiji: Bahá'í Publishing Trust, 1982), pp. 85–6.

29 Letter on behalf of Shoghi Effendi to an individual believer, 24 February 1947. Quoted in a letter from the Universal House of Justice, Department of the Secretariat, 4 September 2005.

30 Shoghi Effendi. *Unfolding Destiny: The Messages from the Guardian of the Bahá'í Faith to the Bahá'í Community of the British Isles* (London: Bahá'í Publishing Trust, 1981), p. 458.

Chapter 2: An Exchange of Views

31   Mason Remey, *Star of the West* (Chicago: Baha'i News Service, vol. 13, 21 March 1922), p. 39

32   J. E. Esslemont, *Bahá'u'lláh and the New Era* (1923; Wilmette: Bahá'í Publishing Committee, 1937 revision, the basis for the many revisons since published), p. 251.

33   George Latimer, *Star of the West* (Chicago: Baha'i News Service, vol. 16, September 1925), p. 550.

34   John Ferraby, *All Things Made New: A Comprehensive Outline of the Bahá'í Faith* (George Allen & Unwin, 1957), pp. 158-9.

35   Hasan Balyuzi, *'Abdu'l-Bahá* (London: George Ronald, 1971), pp. 296-7.

36   John Hatcher, *The Purpose of Physical Reality* (Wilmette: Bahá'í Publishing Trust, 1987), p. 53

37   Paul Lample (comp.), *Bahá'u'lláh's Teachings on Spiritual Reality* (Riviera Beach, FL: Palabra Publications, 1996), p. 71.

38   Keven Brown and Eberhard von Kitzing: *Evolution and Bahá'í Belief: 'Abdu'l-Bahá's Response to Nineteenth-Century Darwinism* (Studies in the Bábi and Bahá'í Religions, vol. 12. Los Angeles, CA: Kalimát Press, 2001), p. xxi.

39   Ibid. p. 117.

40   Ibid. p. xix.

41   'Abdu'l-Bahá, *Paris Talks*, no. 54, p. 221.

42   Ibid. no. 2, p. 9.

43   Ibid. no. 28, p. 102.

44   'Abdu'l-Bahá, *The Promulgation of Universal Peace*, p. 67.

45   Loehle, *On the Shoulders of Giants*, p. 114.

46   Ibid. p. 110.

47   Salman Oskooi, *When Science and Religion Merge: A Modern Case Study* (2009).

      Available at https://bahai-library.com/
      oskooi_disharmony_science_religion

48   Salman Oskooi, *When Science and Religion Merge: A Modern Case Study* (2009), p. 35.

49   Shoghi Effendi, *Directives from the Guardian* (Comp. Gertrude Garrida. New Delhi: Bahá'í Publishing Trust, 1973), pp. 33–4.

50   Letter on behalf of the Universal House of Justice to individual believers, 3 June 1982, in The Universal House of Justice, *Messages from the Universal House of Justice 1963–1986* (Comp. Geoffry W. Marks. Wilmette, IL: Bahá'í Publishing Trust, 1996), no. 330, pp. 545–6.

    Available at: http://bahai-library.com/uhj_infallibility_abdulbaha.

51   'Abdu'l-Bahá, *The Promulgation of Universal Peace*, p. 495.

52   'Abdu'l-Bahá, *The Promulgation of Universal Peace*, p. 427.

53   'Abdu'l-Bahá, *Paris Talks,* no. 15, p. 51.

Chapter 3: Translating, Interpreting, and Investigating

54   Foreword to 'Abdu'l-Bahá, *Some Answered Questions* (2014), pp. xiv–xv.

55   'Abdu'l-Bahá, *The Promulgation of Universal Peace*, p. 86.

56   Letter on behalf of the Universal House of Justice to an individual, 21 February 2016.

    Available at www.bahai.org/library

57   Shoghi Effendi, *The World Order of Bahá'u'lláh* (Wilmette, IL: Bahá'í Publishing Trust), p. 150.

58   Letter on behalf of the Universal House of Justice to an individual, 22 August 1977.

Available at https://www.bahai.org/library/authoritative-texts/compilations/universal-house-of-justice-compilation/

59   Letter on behalf of the Universal House of Justice to an individual, 25 October 1984, in The Universal House of Justice, *Messages from the Universal House of Justice 1963–1986* (Comp. Geoffry W. Marks. Wilmette, IL: Bahá'í Publishing Trust, 1996), pp. 645-6.

    Available at: https://www.bahai.org/library/authoritative-texts/the-universal-house-of-justice/messages/

60   'Abdu'l-Bahá, *Will and Testament Testament* (Wilmette, IL: National Spiritual Assembly of the Bahá'is of the United States, 1944), para. 9.

61   Letter of 25 October 1984, in *Messages from the Universal House of Justice: 1963-1986.*

62   Letter of 9 March 1965, in *Wellspring of Guidance: Messages of the Universal House of Justice 1963-68* (Wilmette, IL: Bahá'i Publishing Trust, 1969), p. 53.

63   Ibid. p. 84.

64   Letter of 27 May 1966, in *Wellspring of Guidance: Messages of the Universal House of Justice 1963-68* (Wilmette, IL: Bahá'i Publishing Trust, 1969), pp. 88-89

65   Ibid.

66   Letter of 21 July 1968 to an individual believer, quoted in *A Compilation on Scholarship* (Bahá'i World Centre, February 1995).

67   Letter of 26 December 1975, quoted in *A Compilation on Scholarship* (Bahá'i World Centre, February 1995).

68   Letter from the Universal House of Justice, Department of the Secretariat, to an individual, 4 September 2005.

69   Letter from the Universal House of Justice, Department of the Secretariat, to an individual, 5 July 2010.

70   Qur'án 36:38–40, 14:33, 21:33, 79:3.

71   'Abdu'l-Bahá, *Some Answered Questions*, no. 7, pp. 28–9.

72   Gary Matthews, *The Challenge of Bahá'u'lláh* (Wilmete, IL: Bahá'i Publishing, 2005), p. 93. [First edition published 1993 by George Ronald]

73   Ibid. p. 95.

74   Ibid. p. 93.

75   Bahman Nadimi, Do the Baha'i Writings on evolution allow for mutation of species within kingdoms but not across kingdoms?', at *Bahá'i Library Online* (2004)

Chapter 4: The Base of the Tree of Life

76   Bahá'u'lláh, *Gleanings from the Writings of Bahá'u'lláh* (Wilmette, IL: Bahá'i Publishing Trust, 2nd ed. 1976), LXXXII, p. 163.

77   David Raup and James Valentine, 'Multiple origins of life' *Proceedings of the National Academy of Sciences USA* (May 1983), Vol 80, pp. 2981-2984.

78   Formation of nucleobases in a Miller–Urey reducing atmosphere. PNAS 2017 ; published ahead of print April 10, 2017

79   Georgia Institute of Technology. (24 January 2019). The helix, of DNA fame, may have arisen with startling ease. *ScienceDaily*.

80   University of Colorado at Boulder. (7 April 2015). New study hints at spontaneous appearance of primordial DNA. *ScienceDaily*.

81   Robert F. Service, 'Researchers may have solved origin-of-life conundrum' *Science* (16 March 2015).

82   University of Tokyo. (2022, March 18). New insight into the possible origins of life: Experiment sheds light on the molecular evolution of RNA. *ScienceDaily*.

83   Jerome CA, Kim HJ, Mojzsis SJ, Benner SA, Biondi E. Catalytic Synthesis of Polyribonucleic Acid on Prebiotic Rock Glasses. *Astrobiology*. 2022 Jun;22(6):629-636. doi: 10.1089/ast.2022.0027. Epub 2022 May 19. PMID: 35588195; PMCID: PMC9233534.

84   William C. White and David C. Culver (eds), *Encyclopedia of Caves* (Academic Press, 2nd ed. online 7 February 2012).

85   Sandi Doughton,'Ice worms: They're real, and they're hot', in *Seattle Times*, 21 February 2006.

86   Jennifer Carpenter, 'Deepest-living land animal found', in *BBC News*, 2 June 2001.

87   National Aeronautics and Space Administration (NASA). 'Earth microbes on the moon', in *NASA Space Science News*, 1 September 1998.

88   Emma Brennand, 'Tardigrades: Water bears in space', in *BBC Nature*, 17 May 2011.

89   'Abdu'l-Bahá, *Abdul-Baha on Divine Philosophy* (Comp. I. F. Chamberlain. Boston, MA: Tudor Press, 1918), pp. 114–15.

90   University College London, 'Energy revolution key to complex life: Depends on mitochondria, cells' tiny power stations', in *ScienceDaily*, 21 October 2010.

91   Carl Woese, 'On the evolution of cells', in *Proceedings of the National*

Academy of Sciences USA, vol. 99, no. 13 (25 June 2002), pp. 8742–8747.

92   Ibid.

93   Carl Woese, 'A new biology for a new century', in *Microbiology and Molecular Biology*, vol. 68, no. 2 (rev. June 2004), pp. 173–186.

94   Woese, 'On the evolution of cells', op. cit.

95   D. Wu, M. Wu, A. Halpern et al., 'Stalking the fourth domain in metagenomic data', in PLOS ONE, 18 March 2011;

See also 'A new domain of life: Plenty more bugs in the sea', in *The Economist*, 24 March 2011.

96   Rachel Moeller, 'New theory of cell evolution rejects single-ancestor doctrine', in *Scientific American*, 19 June 2002.

97   W. Ford Doolittle and Eric Bapteste, 'Pattern pluralism and the Tree of Life hypothesis', in *Proceedings of the National Academy of Sciences USA*, vol. 104, no. 7 (13 February 2007), pp. 2043–2049.

98   W. Ford Doolittle, 'Uprooting the tree of life', in *Scientific American*, vol. 282, no. 2 (February 2000), pp. 90–95.

99   Eugene McCarthy, *On the Origins of New Forms of Life: A New Theory* (2008), p. 1. Available at: http://www.macroevolution.net/support-files/forms_of_life.pdf.

100   'Charles Darwin's tree of life is "wrong and misleading"', claim scientists', in *The Telegraph*, 22 January 2009.

101   Ian Sample, 'Evolution: Charles Darwin was wrong about the tree of life', in *The Guardian*, 21 January 2009.

102   V. Kunin, L. Goldovsky, N. Darzentas and C. A. Ouzounis, 'The net of life: Reconstructing the microbial phylogenetic network', in *Genome Research*, vol. 15 (2005), pp. 954–959.

103   Barton et al., *Evolution* (2007), p. 131-132.

104   National Center for Science Education, 'Malcolm Gordon', 23 October 2008. https://ncse.ngo/malcolm-gordon-1

105   B. Marin, E. Nowack and M. Melkonian, 'A plastid in the making: Evidence for a second primary endosymbiosis', in *Protist*, vol. 156 (12 October 2005), pp. 425–432.

106   N. Okamoto and I. Inouye, 'A secondary symbiosis in progress?', in *Science*, vol. 310, no. 5746 (14 October 2005), p. 287.

107   C. J. Howe, A. C. Barbrook, R. E. R. Nisbet et al., 'The origin of plastids', in *Philosophical Transactions of the Royal Society London, Series B, Biological Sciences*, vol. 363, no. 1504 (27 August 2008), pp. 2675–2685.

108   Fabien, Burki, Kamran Shalchian-Tabrizi and Jan Pawlowski, 'Phylogenomics reveals a new "megagroup" including most photosynthetic eukaryotes', in *Biology Letters*, vol. 4, no. 4 (23 August 2008), pp. 366–369.

109   Cheong Xin Chan, Jefferson Gross, Hawan Su Yoon and Debashish Bhattacharya, 'Plastid origin and evolution: New models provide insights into old problems', in *Plant Physiology*, vol. 155, no. 4 (April 2011), pp. 1552–1560.

110   S. M. Miller, 'Volvox, chlamydomonas, and the evolution of multicellularity', in *Nature Education*, vol. 3, no. 9 (2010), p. 65.

111   Public Library of Science, 'Model explains rapid transition toward division of labor in biological evolution', in *ScienceDaily*, 10 June 2010.

112   William C. Ratcliff, R. Ford Denisona, Mark Borrelloa and Michael Travisano, 'Experimental evolution of multicellularity', in *Proceedings of the National Academy of Sciences USA*, vol. 109 (2011), no. 5, pp. 1595–1600.

113   Richard K. Grosberg and Richard R. Strathmann, 'The evolution of multicellularity: A minor major transition?', in *Annual Review of Ecology, Evolution, and Systematics*, 10 August 2007, pp. 621–654.

114   Stuart A. Newman, 'Physico-genetic determinants in the evolution of development', in *Science*, vol. 338, no. 6104 (12 October 2012), pp. 217–219.

115   Ibid.

116   Elsevier, 'New theory, embryo geometry, proposes explanation for how vertebrates evolved', in *ScienceDaily*, 31 August 2016.

117   David B. Edelman, Mark McMenamin, Peter Sheesley and Stuart Pivar, 'Origin of the vertebrate body plan via mechanically biased conservation of regular geometrical patterns in the structure of the blastula', in *Progress in Biophysics and Molecular Biology*, vol. 121, issue 3 (September 2016), pp. 212–224.

118   Ibid.

119   Guillaume Balavoine and André Adoutte, 'The segmented Urbilat-

eria: A testable scenario', in *Integrative and Comparative Biology*, vol. 43 (2003), no. 1, pp. 137–147.

120   Leonid L. Moroz, 'On the independent origins of complex brains and neurons', in *Brain, Behavior and Evolution*, vol. 74, no. 3 December 2009), pp. 177–190.

121   Malcolm Gordon, 'The Concept of Monophyly: A Speculative Essay', in *Biology and Philosophy*, vol. 14 (1999), pp. 331-348.

122   Ibid. p. 335

123   Periannan Senapathy, *Independent Birth of Organisms* (Madison, WI: Genome Press, 2004).

124   Ibid. p. 10.

125   Christian Schwabe, *The Genomic Potential Hypothesis* (Georgetown, TX: Landes Bioscience, 2001).

126   Christian Schwabe, 'Embryotic evolution: An ancient question, a new answer', in *Cell Cycle*, vol. 7 (2008), pp. 1503–6.

Chapter 5: Convergence

127   McCarthy, *On the Origins of New Forms of Life*, pp. 255–261.

128   Eberhard von Kitzing, 'The Origin of Complex Order in Biology' in *Evolution and Bahá'í Belief*, (Los Angeles: CA, Kalimat Press, 2001), p. 234.

129   'Unfinished business: Charles Darwin's ideas have spread widely, but his revolution is not yet complete', in *The Economist*, 5 February 2009.

130   Conway Morris, *Life's Solution: Inevitable Humans in a Lonely Universe*, p. xii.

131   Jonathan B. Losos. *Improbable Destinies: Fate, Chance, and the Future of Evolution* (New York: Riverhead Books, 2017), preface, p. 2.

132   Ibid. p. 43.

133   G. J. Vermeij, 'Historical contingency and the purported uniqueness of evolutionary innovations', in *Proceedings of the National Academy of Sciences USA*, vol. 103 (2006), pp. 1804–1809.

134   Carol Kaesuk Yoon, 'Is evolution truly random?', in *The New York Times, Science*, 11 November 2003.

135 University of Chicago Medical Center, 'Flatfish fossils fill in evolutionary missing link', in *ScienceDaily*, 10 July 2008.

136 Matthew Campbell, et al. 'Are flatfishes (Pleuronectiformes) monophyletic?.' *Molecular phylogenetics and evolution* vol. 69,3 (2013);

Carl Zimmer, 'The Evolution of Extraordinary Eyes: The Cases of Flatfishes and Stalk-Eyed Flies' in *Evolution: Education and Outreach* (16 October 2008), vol. 1, issue 4.

137 H. Utsuno, T. Asami, T. J. van Dooren and E. Gittenberger. 'Internal selection against the evolution of left–right reversal', in *Evolution*, vol. 65 (2011), pp. 2399–2411.

138 Masaki Hoo, Takahiro Asami and Michio Hori, 'Right-handed snakes: convergent evolution of asymmetry for functional specialization', in *Biology Letters*, vol. 3, issue 2 (22 April 2007), pp. 169-U2.

139 R. A. Norberg, 'Occurrence and independent evolution of bilateral ear asymmetry in owls and implications on owl taxonomy', in *Philosophical Transactions of the Royal Society, Biological Sciences*, No. 280 (31 August 1977), pp. 375–408.

140 Thewissen Wursig, *Encyclopedia of marine mammals* (London: Academic Press, 2002).

141 Acoustical Society of America (ASA), 'Bats, whales, and bio-sonar: New findings about whales' foraging behavior reveal surprising evolutionary convergence', in *ScienceDaily*, 8 May 2012.

142 University of London (Queen Mary), 'Genetic similarities between bats and dolphins discovered', in *ScienceDaily*, 4 September 2013.

143 University of Strathclyde, 'Dolphin hearing system component found in insects', in *ScienceDaily*, 13 December 2012.

144 Lewis Wolpert, 'Development of the asymmetric human', in *European Review*, vol. 13, supp. no. 2 (2005), pp. 97–103.

145 Michael Levin, 'Left-right asymmetry in embryonic development: A comprehensive review', in *Mechanisms of Development*, no.122 (2005), pp. 3–25.

146 N. A. Brown and L. Wolpert, 'The development of handedness in left/right asymmetry', in *Development*, vol. 109, no. 1 (May 1990), pp. 1–9.

147 Tufts University, 'Human cells, worms, frogs and plants share mecha-

nism for asymmetrical patterning: tubulin proteins', in *ScienceDaily*, 16 July 2012.

148   A. Davison, J. S. McDowell, J. M. Holden, J. M. et al. 'Formin is associated with leftright asymmetry in the pond snail and the frog', in *Current Biology*, vol. 26, issue 5 (7 March 2016), pp. 654–660.

149   'Abdu'l-Bahá, *Some Answered Questions*, no. 49, para. 3, p. 220.

150   Ibid. no. 49, para. 5, p. 222.

151   Gregory D. Amoutzias, David L.Robertson, Stephen G. Oliver and Erich Bornberg- Bauer, 'Convergent evolution of gene networks by single-gene duplications in higher eukaryotes', in *EMBO Reports*, vol. 5 (2004), pp. 274– 279.

152   Michel Tassetto, Alexis Maizel, Joana Osorio and Alain Joliot, 'Plant and animal homeodomains use convergent mechanisms for intercellular transfer', in *EMBO Reports*, vol. 6 (2005), pp. 885–890.

153   Yang Liu, James A. Cotton, Bin Shen et al., 'Convergent sequence evolution between echolocating bats and dolphins', in *Current Biology*, vol. 20, issue 2 (26 January 2010), pp. R53–R54.

154   Melissa Ilardo, Markus Meringer, et al. 'Adaptive Properties of the Genetically Encoded Amino Acid Alphabet Are Inherited from Its Subsets.' *Scientific Reports* **9**, 12468 (2019)

       Scripps Research Institute. "A chemical clue to how life started on Earth: Every living thing stems from the same limited set of 20 amino acids, and now scientists may know why." *ScienceDaily*, 1 August 2019.

155   Mary Jane West-Eberhard, 'Development plasticity and the origin of species differences', in *Proceedings of the National Academy of Sciences* (PNAS), vol. 102 (3 May 2005), pp. 6543–6549.

156   Jonathan B. Losos. *Improbable Destinies: Fate, Chance, and the Future of Evolution* (New York: Riverhead Books, 2017), pp. 118-119.

157   Julian Hume and David Martill. 'Repeated evolution of flightlessness in Dryolimnas rails after extinction and recolonization on Aldabra' in *Zoological Journal of the Linnean Society* (08 May 2019), vol. 186, issue 3, pp. 666-672.

158   Simon Conway Morris. *Runes of Evolution: How the Universe Became Self-Aware* (Templeton Press; 2015), p. 32.

159   Ibid. p. 33.

160 Canterbury Museum. "New Zealand's ancient monster penguins had northern hemisphere doppelgangers." in *ScienceDaily*, 30 June 2020.

161 Charles Darwin, *On the Origin of Species* (1859; London: John Murray, 6th ed. 1872), Chapter 4.

162 P. R. Grant, R. B. Grant, J. A. Markert et al., 'Convergent evolution of Darwin's finches caused by introgressive hybridization and selection', in *Evolution*, vol. 7 (2004), pp. 1588–1599.

163 Carl R Woese (June 2004). 'A New Biology for a New Century' *Microbiology and Molecular Biology Reviews* 68 2:173-186.

164 'Abdu'l-Bahá, *Some Answered Questions*, no. 49, p. 223.

Chapter 6: Rapid Evolution

165 G. J. Vermeij, 'Historical contingency and the purported uniqueness of evolutionary innovations', in *Proceedings of the National Academy of Sciences USA*, vol. 103 (2006), pp. 1804–1809.

166 BioMed Central Limited. "Advantages of living in the dark: Multiple evolution events of 'blind' cavefish." *ScienceDaily* (22 January 2012).

167 Jonathan B. Losos. *Improbable Destinies: Fate, Chance, and the Future of Evolution* (New York: Riverhead Books, 2017), p. 82

168 Ibid. p. 85.

169 Ibid. p. 86.

170 Ibid. p. 191.

171 Ibid. p. 196.

172 Ibid. p. 208.

173 Ibid. p. 234.

174 Harvard University. "Evolution in vertebrates", *ScienceDaily* (31 January 2019).

175 Richard Lenski, 'Evolution in Action', *Microbe* (2011), vol. 6, no. 1, p. 32.

176 Losos. *Improbable Destinies* (2017), p. 296.

177 Neil deGrasse Tyson, 'Is Anyone Out There Like Us?', *Natural History* (September 1996).

178 Losos. *Improbable Destinies* (2017), p. 368.

Chapter 7: The Human Essence

179   'Abdu'l-Bahá, *Some Answered Questions*, no. 49, p. 220.

180   Robert Wright, 'One world, under God', in *The Atlantic*, April 2009.

181   'No words to describe monkeys' play', *BBC News*, 9 May 2003.

182   Woese, 'A new biology for a new century', in *Microbiology and Molecular Biology*, vol. 68, no. 2 (rev. June 2004), pp. 173–186.

183   McCarthy, *On the Origins of New Forms of Life*, p. 186.

184   Brown University, 'Marsupial wolf or Tasmanian tiger? Extinct Australian thylacine was more cat than dog, researchers find', in *ScienceDaily*, 4 May 2011.

185   Quoted in McCarthy, On the Origins of New Forms of Life, p. 26.

186   Coyne, *Why Evolution is True*, p. 186.

187   'Abdu'l-Bahá, *Some Answered Questions*, no. 50, p. 225.

188   Brown, *Evolution and Bahá'í Belief*, p. 93.

189   'Abdu'l-Bahá, *Some Answered Questions*, no. 80, p. 323.

190   Bahá'u'lláh, *Gleanings from the Writings of Bahá*'u'lláh, XXVII, p. 65.

191   Richard Dawkins (1976), *The Selfish Gene*, Introduction (Oxford: Oxford University Press, 2006 edition).

192   Simon Conway Morris, 'Darwin was right, up to a point', in *The Guardian*, 12 February 2009.

193   Bobby Henderson, Open Letter to Kansas School Board, 2005.

194   Discovery Institute, Center for the Renewal of Science & Culture, *The Wedge* (Seattle, WA, 1999). Shared online: http://www.antievolution.org/features/wedge.pdf.

195   NOVA (US science TV series), *Judgment Day: Intelligent Design on Trial*, aired 13 November 2007 on PBS.

196   S. C. Meyer, 'The origin of biological information and the higher taxonomic categories', in *Proceedings of the Biological Society of Washington*, vol. 117 (2004), no. 2, pp. 213–239.

197   Bahá'u'lláh, Lawḥ-i-Ḥikmat (Tablet of Wisdom), in *Tablets of Bahá*'u'lláh *Revealed after the Kitáb-i-Aqda*s (Wilmette, IL: Bahá'í Publishing Trust. 1988 edition), p. 141.

198   Ibid. pp. 141–144.

199 'Abdu'l-Bahá, *Selections from the Writings of 'Abdu'l-Bahá* (Haifa: Bahá'í World Centre, 1978), no. 30, p. 61.

200 'Abdu'l-Bahá, Tablet to August Forel, in *The Bahá'í World*, vol. XV (Haifa: Bahá'í World Centre, 1976), p. 18; also in John Paul Vader, *For the Good of Mankind: August Forel and the Bahá'í Faith* (Oxford: George Ronald, 1984), p. 76.

201 Bahá'u'lláh, *Gleanings from the Writings of Bahá'u'lláh*, XXVII, pp. 65–66.

202 Ibid. LXXVIII, pp. 150–151.

203 Shoghi Effendi, *The World Order of Bahá'u'lláh* (1938; Wilmette, IL: Bahá'í Publishing Trust, 2nd rev. ed. 1974), p. 115.

Chapter 8: Human Fossils

204 'Abdu'l-Bahá, *The Promulgation of Universal Peace*, p. 507.

205 Cleveland Museum of Natural History, '3.6 million-year-old relative of "Lucy" discovered: Early hominid skeleton confirms human-like walking is ancient', in *ScienceDaily*, 21 June 2010.

206 American Association for the Advancement of Science, 'Before "Lucy," there was "Ardi": First major analysis of early hominid published in science', in *ScienceDaily*, 1 October 2009.

207 Jamie Shreeve, 'Oldest skeleton of human ancestor found', in *National Geographic Magazine*, 1 October 2009.

208 Duke University, 'Bipedal humans came down from the trees, not up from the ground' in *ScienceDaily*, 11 August 2009.

209 Terry A. Vaughan, James M. Ryan and Nicholas J. Czaplewski, *Mammalogy*, (Sudbury, MA: Jones & Bartlett Learning, 5th ed. 2010), p. 191.

210 Roger Lewin, *Human Evolution: An Illustrated Introduction* (Oxford: Wiley-Blackwell, 2004), p. 103.

211 Field Museum, 'New statistical model moves human evolution back three million years', in *ScienceDaily*, 5 November 2010.

212 Robert Ballard, *The Eternal Darkness* (Princeton: Princeton University Press, 2002).

213 University of Utah, 'Six million years of savanna: Grasslands,

wooded grasslands accompanied human evolution', in *ScienceDaily*. 3 August 2011.

214   Sally McBrearty and Nina G. Jablonski, 'First fossil chimpanzee', in *Nature*, no. 437 (1 September 2005), pp. 105–108.

215   A. E. Lebatard, D. L. Bourlès, P. Duringer et al., 'Cosmogenic nuclide dating of Sahelanthropus tchadensis and Australopithecus bahrelghazali: Mio-Pliocene hominids from Chad', in *Proceedings of the National Academy of Sciences USA*, no. 105 (2008), p. 3226; Jamie Shreeve, 'The evolutionary road', in *National Geographic Magazine* (July 2010), p. 60.

216   'Evolution's human and chimp twist', *BBC News* (18 May 2006).

217   M. Ventura, C. R. Catacchio, C. Alkan et al., 'Gorilla genome structural variation reveals evolutionary parallelisms with chimpanzee', in *Genome Research*, vol. 21 (2011), pp. 1640–1649.

218   A. E. Lebatard et al., 'Cosmogenic nuclide dating of Sahelanthropus tchadensis...', op. cit.

219   Rex Dalton, 'Oldest gorilla ages our joint ancestor', in *Nature*, no. 448 (23 August 2007), pp. 844–845.

220   T. D. White, G. Suwa and C. O. Lovejoy, 'Response to comment on the paleobiology and classification of Ardipithecus ramidus', in *Science*, No. 328 (2010), p. 1105.

221   Broad Institute of MIT and Harvard, 'Human and chimp genomes reveal new twist on origin of species', in *ScienceDaily*, 18 May 2006.

222   Bernard Wood, 'Palaeoanthropology: Hominid revelations from Chad', in *Nature*, no. 418 (11 July 2002), pp. 133–135.

223   Uppsala University, 'Fossil footprints challenge established theories of human evolution', in *ScienceDaily*, 31 August 2017.

224   'Possible hominin footprints from the late Miocene (c. 5.7 Ma) of Crete?', in *Proceedings of the Geologists' Association*, vol. 128, issues 5–6 (October 2017), pp. 697–710.

225   University of Toronto, '7.2-million-year-old pre-human remains found in the Balkans: New hypothesis about the origin of humankind suggests oldest hominin lived in Europe', in *ScienceDaily*, 23 May 2017.

226   Carol V. Ward et al, 'A late Miocene hominid partial pelvis from Hungary', *Journal of Human Evolution* vol. 136 (November 2019).

227   Stony Brook Medicine (2013, December 4). 'Early tree-dwelling bipedal human ancestor was similar to ancient apes and "Lucy" but not living apes', in *ScienceDaily*, 4 December 2013.

228   Florida Museum of Natural History. "Oldest-known ancestor of modern primates may have come from North America, not Asia." ScienceDaily. ScienceDaily, 29 November 2018.

229   University of Pittsburgh, 'Humans more related to orangutans than chimps, study suggests', in *ScienceDaily*, 18 June 2009.

230   Universitat Autònoma de Barcelona, 'New hominid 12 million years old found in Spain, with "modern" facial features', in *ScienceDaily*, 2 June 2009.

231   John R. Grehan and Jeffrey H. Schwartz (2009). 'Evolution of the second orangutan: Phylogeny and biogeography of hominid origins', in *Journal of Biogeography*, Special Paper (2009).

232   A. Hobolth, J. Y. Dutheil, J. Hawks et al., 'Incomplete lineage sorting patterns among human, chimpanzee, and orangutan suggest recent orangutan speciation and widespread selection', in *Genome Research*, vol. 21(2011), pp. 349–356.

233   Jon Cohen, 'Orangutan genome full of surprises', in *Science Now, Evolution*, 26 January 2011.

234   Wellcome Trust Sanger Institute, 'What have we got in common with a gorilla? Insight into human evolution from gorilla genome sequence', in *ScienceDaily*, 7 March 2012.

Chapter 9: Genomes

235   Wellcome Trust Sanger Institute, 'We are all mutants: First direct whole-genome measure of human mutation predicts 60 new mutations in each of us', in *ScienceDaily*, 13 June 2011.

236   Abdu'l-Bahá, *Some Answered Questions*, no. 49, p. 222.

237   Royal Botanic Gardens, Kew, 'Rare Japanese plant has largest genome known to science', in *ScienceDaily*, 7 October 2010.

238   Mouse Genome Analysis Group, 'Initial sequencing and comparative analysis of the mouse genome', in *Nature*, no. 420 (5 December 2002), pp. 520–62.

239   Yale University, 'Not "junk DNA" after all: Tiny RNAs play big role controlling genes', in *ScienceDaily*, 26 October 2007.

240   Flanders Institute for Biotechnology (VIB), 'Saved by junk DNA: Vital role in the evolution of human genome', in *ScienceDaily* 30 May 2009,

241   'The new world of DNA: A long-term effort to catalogue all the bits of the human genome that do something has released its results', in *The Economist*, 5 September 2012.

242   'The nature of man: Large-scale genetic studies are throwing light on what makes humans human', in *The Economist*, 5 September 2012.

243   Nicholas Wade, 'Study detects recent instance of human evolution', in *The New York Times*, 10 December 2006;

Sarah A .Tishkoff, Floyd A. Reed, Alessia Ranciaro et al., 'Convergent adaptation of human lactase persistence in Africa and Europe', in *Nature Genetics*, vol. 39 (2007), no. 1, pp. 31–40.

244   N. S. Enattah, A. Trudeau, V. Pimenoff et al., 'Evidence of still-ongoing convergence evolution of the lactase persistence T-13910 alleles in humans', in *American Journal of Human Genetics*, vol. 81 (2007), no. 3, pp. 615–625.

245   R.A. Raff, C.R. Marshall and J. M. Turbeville, 'Using DNA sequences to unravel the Cambrian radiation of the animal phyla', in *Annual Review of Ecology and Systematics*, vol. 25 (1994), pp. 351–375.

246   Gavin C. Conant and Andreas Wagner, 'Convergent evolution of gene circuits', in *Nature Genetics*, vol. 34 (2003), pp. 264–266.

247   E. J. Vowles and W. Amos, (2004) 'Evidence for widespread convergent evolution around human microsatellites', in *Public Library of Science (PLoS) Biology*, vol. 2 (2004), no. 8, e199.

248   I. Maeso, S. W. Roy and M. Irimia, 'Widespread recurrent evolution of genomic features', in *Genome Biology and Evolution*, vol. 4 (2012), no. 4, pp. 486–500.

249   Duke University, 'Roots of language in human and bird biology: Genes activated for human speech similar to ones used by singing songbirds', in *ScienceDaily* 14 February 2013.

250   Marie Manceau, et al. 'Convergence in pigmentation at multiple levels: mutations, genes and function' in *Philosophical Transactions of the Royal Society B* (27 August 2010), vol. 365, issue 1552.

251 P. A. Christin, D. M. Weinreich and G. Besnard, 'Causes and evolutionary significance of genetic convergence', in *Trends in Genetics*, vol. 26 (2010), no. 9, pp. 400–405.

252 Parker, J., Tsagkogeorga, G., Cotton, J. *et al.* Genome-wide signatures of convergent evolution in echolocating mammals. *Nature* **502**, 228–231 (2013) doi:10.1038/nature12511

253 Howard Hughes Medical Institute, 'Analysis of stickleback genome sequence catches evolution in action: Reuse of key genes is common theme', in *ScienceDaily*, 4 April 2012.

254 Public Library of Science (PLoS), 'Distantly related fish find same evolutionary solution to dark water', in *ScienceDaily*, 11 April 2017.

255 Max Planck Institute, 'Repeat act: Parallel selection tweaks many of the same genes to make big and heavy mice', in *ScienceDaily*, 8 May 2012.

256 Cornell University, 'Same adaptations evolve across different insects', in *ScienceDaily*, 24 July 2012.

257 Princeton University, 'Far from random, evolution follows a predictable genetic pattern', in *ScienceDaily*, 25 October 2012.

258 Washington University in St Louis, 'How repeatable is evolutionary history? "Weakness" in clover genome biases species to evolve same trait', in *ScienceDaily*, 23 June 2014.

259 University of Wisconsin (Madison), 'The shocking truth about electric fish: Genomic basis for the convergent evolution of electric organs', in *ScienceDaily*, 26 June 2014.

260 University of California (Santa Barbara), 'Let there be light: Evolution of complex bioluminescent traits may be predictable', in *ScienceDaily*, 21 October 2014.

261 Buller, Andrew R, and Craig A Townsend. 'Intrinsic evolutionary constraints on protease structure, enzyme acylation, and the identity of the catalytic triad.' *Proceedings of the National Academy of Sciences of the United States of America* vol. 110,8 (2013): E653-61. doi:10.1073/pnas.1221050110

262 P. A. Christin, G. Besnard, E. J. Edwards and N. Salamin, 'Effect of genetic convergence on phylogenetic inference', in *Molecular Phylogenetics and Evolution*, vol. 62 (2012), no. 3, pp. 921–927.

263 Parker, J., Tsagkogeorga, G., Cotton, J. *et al.* Genome-wide signatures

of convergent evolution in echolocating mammals. *Nature* **502,** 228–231 (2013) doi:10.1038/nature12511

264 Castoe TA, *et al.* 'Adaptive molecular convergence: Molecular evolution versus molecular phylogenetics.' *Commun Integr Biol* (January 2010); **3**(1):67-9.

265 Foote, A., Liu, Y., Thomas, G. *et al.* Convergent evolution of the genomes of marine mammals. *Nat Genet* **47,** 272–275 (2015) doi:10.1038/ng.3198

266 Chikina M, *et al.* 'Hundreds of Genes Experienced Convergent Shifts in Selective Pressure in Marine Mammals.' *Mol Biol Evol* (September 2016); **33**(9):2182-92. Epub 21 Jun 2016.

267 Partha R, *et al.* 'Subterranean mammals show convergent regression in ocular genes and enhancers, along with adaptation to running.' *eLife* (16 October 2017), DOI: 10.7554/eLife.25884.

268 Hu, Yibo et al. "Comparative genomics reveals convergent evolution between the bamboo-eating giant and red pandas." *Proceedings of the National Academy of Sciences of the United States of America* vol. 114,5 (2017): 1081-1086. doi:10.1073/pnas.1613870114

269 Merenyi Z, et al. 'Unmatched level of molecular convergence among deeply divergent complex multicellular fungi' *bioRxiv* 549758; doi: https://doi.org/10.1101/549758

270 Charles Y. Feigin et al. 'Widespread cis-regulatory convergence between the extinct Tasmanian tiger and gray wolf', *Genome Research* (2019).

271 Okinawa Institute of Science and Technology (OIST) Graduate University. "Our closest wormy cousins: About 70% of our genes trace their ancestry back to the acorn worm." *ScienceDaily* (18 November 2015)

272 NOVA, *Cracking the Code of Life,* aired 17 April 2001 on PBS.

273 D. Dryden, A. Thomson and J. White, 'How much of protein sequence space has been explored by life on earth?', in *Journal of the Royal Society Interface*, vol. 5 (208) no. 25, pp. 953–956.

274 T. Anzai, T. Shiina, N. Kimura et al., 'Comparative sequencing of human and chimpanzee MHC class I regions unveils insertions/deletions as the major path to genomic divergence', in *Proceedings of the National Academy of Sciences USA*, vol. 100 (2003), no. 13, pp. 7708–7713.

275 Coyne, *Why Evolution is True*, p. 230.

276 Jonathan Marks, 'Chronicle of higher education' (2000), RP in W. Haviland, H. Prins, D. Walrath and B. McBride: *The Essence of Anthropology* (Belmont, CA: Wadsworth, 2010), p. 39.

277 Jon Cohen, 'Relative differences: The myth of 1%', in *Science*, vol. 316 (2007), no. 5833, p. 1836.

278 Stony Brook University Medical Center, 'Upright walking began 6 million years ago, thigh bone comparison suggests', in *ScienceDaily*, 21 March 2008.

279 Kansas State University, 'Chromosomes' big picture: Similarities found in genomes across multiple species; Platypus still out of place', in *ScienceDaily*, 11 July 2011.

280 University of Texas (Arlington), 'Evolutionary surprise: Eight percent of human genetic material comes from a virus', in *ScienceDaily*, 8 January 2010.

281 Texas A&M University (Agricultural Communications), 'Researchers discover that sheep need retroviruses for reproduction', in *Science-Daily*, 11 September 2006.

282 A. Dupressoir, C. Vernochet, O. Bawa et al., 'Syncytin-A knockout mice demonstrate the critical role in placentation of a fusogenic, endogenous retrovirus-derived, envelope gene', in *Proceedings of the National Academy of Sciences USA*, vol. 106 (2009), no. 29, pp. 12127–12132.

283 M. T. Romanish, W. M. Lock, L. N. van de Lagemaat et al., 'Repeated recruitment of LTR retrotransposons as promoters by the anti-apoptotic locus NAIP during mammalian evolution', in *PLoS Genetics,* vol. 3 (2007), no. 1, e10.

284 American Society for Microbiology, 'Upending conventional wisdom, certain virus families are ancient', in *ScienceDaily*, 17 November 2010.

285 Yale University, 'Invasion of genomic parasites triggered modern mammalian pregnancy, study finds', in *ScienceDaily*, 26 September 2011.

286 Stanford University Medical Center, 'Ancient viral molecules essential for human development', in *ScienceDaily*, 23 November 2015.

287 R. Belshaw, A. L. A. Dawson, J. Woolven-Allen et al., 'Genome-wide screening reveals high levels of insertional polymorphism in the human endogenous retrovirus family HERV-K(HML2): Implications

for present-day activity', in *Journal of Virology*, vol. 79 (2005), no. 19, pp. 12507–12514.

288 Ecole Polytechnique Fédérale de Lausanne, 'Biologists wake dormant viruses and uncover mechanism for survival', in *ScienceDaily*, 15 January 2010;

University of British Columbia, ' "Start/stop switch" for retroviruses found', in *ScienceDaily*, 10 April 2010.

289 M. Ventura, *et al.*, 'Gorilla genome structural variation reveals evolutionary parallelisms with chimpanzee', in *Genome Research*, vol. 21 (2011), pp. 1640–1649;

C. T. Yohn, *et al.*, 'Lineage-specific expansions of retroviral insertions within the genomes of African great apes but not humans and orangutans', in *PLoS Biology*, vol.3 (2005), no.4, e110;

M. Barbulescu, *et al.*, 'A HERV-K provirus in chimpanzees, bonobos and gorillas, but not humans', in *Current Biology*, vol. 11 (2001), pp. 10779–10783.

290 Yohn, Jiang, McGrath et al., 'Lineage-specific expansions...', op. cit.

291 R. A. Weiss, 'The discovery of endogenous retroviruses', in *Retrovirology*, vol. 3 (2006), p. 67.

292 K. Burns and J. D. Boeke, 'Human transposon tectonics', in *Cell*, vol. 149 (2012), no. 4, pp. 720–752;

S. E. Devine and J. D. Boeke, 'Integration of the yeast retrotransposon Ty1 is targeted to regions upstream of genes transcribed by RNA polymerase III', in *Genes and Development*, vol. 10 (1996), no. 5, pp. 620–633;

E. C. Bolton and J. D. Boeke, 'Transcriptional interactions between yeast tRNA genes, flanking genes and Ty elements: a genomic point of view', in *Genome Research*, vol. 13 (2003), no. 3, pp. 254–263.

293 R. S. Mitchell, B. F. Beitze, A. R. Schroder et al., 'Retroviral DNA integration: ASLV, HIV, and MLV show distinct target site preferences', in *PLoS Biology*, vol. 2 (2004), no. 3, e234.

294 V. M. Sarich and A. C. Wilson, 'Immunological time scale for hominid evolution', in *Science*, vol. 158 (1967), no. 3805, pp. 1200–1203.

295 N. Bannert and R. Kurth, 'Retroelements and the human genome: New perspectives on an old relation', in *Proceedings of the National*

Academy of Sciences of the USA, vol. 101 (2004), supplement 2, pp. 14572–14579.

296   A. Crisp, C. Boschetti, M. Perry et al., 'Expression of multiple horizontally acquired genes is a hallmark of both vertebrate and invertebrate genomes', in *Genome Biology*, vol. 16 (2015), no. 50.

297   University of Wisconsin-Madison. "Yeasts reach across tree of life to domesticate suite of bacterial genes." *ScienceDaily*, 21 February 2019.

298   CBC News, 'Leaf-like sea slug feeds on light', 22 January 2012.

299   National Science Foundation, 'How to thrive in battery acid and among toxic metals', in *ScienceDaily*, 7 March 2013.

300   Northwestern University, 'Gonorrhea acquires a piece of human DNA: First evidence of gene transfer from human host to bacterial pathogen', in *ScienceDaily*, 14 February 2011.

301   BioMed Central, 'Parasitic plants steal genes from their hosts', in *ScienceDaily*, 7 June 2012.

302   Vanderbilt University, 'Discovery of jumping gene cluster tangles tree of life', in *ScienceDaily*, 5 February 2011.

303   Max Planck Institute of Molecular Plant Physiology, 'Genetic information migrates from plant to plant', in *ScienceDaily*, 1 February 2012.

304   University at Buffalo, 'Viruses: More survival tricks than previously thought', in *ScienceDaily*, 5 March 2013.

305   Frank Ryan, *Virolution* (London: Harper Collins, 2009).

306   Boston University, 'DNA shape is constrained by evolution: Structural approach to exploring DNA', in *ScienceDaily*, 19 March 2009.

307   S. Parker, L. Hansen, H. Abaan, T. Tullius and E. Margulies, 'Local DNA topography correlates with functional noncoding regions of the human genome', in *Science*, vol. 324 (2009), issue 5925, pp. 389–392.

308   Rockefeller University, 'New nucleotide in DNA could revolutionize epigenetics', in *ScienceDaily*, 17 April 2009.

309   IDIBELL–Bellvitge Biomedical Research Institute, 'Sixth DNA base discovered?', in *ScienceDaily*, 4 May 2015.

310   James Shapiro, *Evolution: A View from the 21st Century* (New Jersey: FT Press Science, 2011), p. 49.

311   Ibid. p. 4.

312   Ibid. p. 6.

313   University of Gothenburg, 'Appearance not always enough to identify species', in *ScienceDaily*, 21 January 2011.

314   BioMed Central, 'Giraffes and frogs provide more evidence of new species hidden in plain sight', in *ScienceDaily*, 2 January 2008.

315   Ibid.

316   European Molecular Biology Laboratory (EMBL), 'From worm muscle to spinal discs: An evolutionary surprise', in *ScienceDaily*, 12 September 2014.

317   University College London. 'Simple marine worms distantly related to humans', in *ScienceDaily*, 10 February 2011.

318   Marine Biological Laboratory, 'Evolutionary surprise: Developmental "scaffold" for vertebrate brain found in brainless marine worm', in *ScienceDaily*, 14 March 2012;

'Simple worms are closest brainiacs', in *New Scientist, Life*, 14 March 2012.

319   Ohio Supercomputer Center, 'Rare deep-sea starfish stuck in juvenile body plan', in *ScienceDaily*, 2 May 2011.

320   Uni Research,'Where does our head come from? Brainless sea anemone sheds new light on the evolutionary origin of the head', in *ScienceDaily*, 20 February 2013.

321   Stowers Institute for Medical Research, 'How an ancient vertebrate uses familiar tools to build a strange-looking head', in *ScienceDaily*, 14 September 2014.

322   Okinawa Institute of Science and Technology (OIST) Graduate University, 'Worm genomes reveal a link between ourselves and our distant relatives: Decoding two worm genomes provides new insights into genetic similarities between distantly related animal groups', in *ScienceDaily*, 4 December 2017.

323   See http://news.richmond.edu/releases/archives/aug05/Hills.html.

324   Merenyi Z, *et al*. 'Unmatched level of molecular convergence among deeply divergent complex multicellular fungi' *bioRxiv* (14 February 2019) **doi**: https://doi.org/10.1101/549758

Chapter 10: Human Potential

325 'Abdu'l-Bahá, *The Promulgation of Universal Peace*, p. 507.

326 'Abdu'l-Bahá, *Some Answered Questions*, no. 46, pp. 205–206.

327 Ibid. no. 52, p. 232–233.

328 David Bjorklund, *Why Youth is Not Wasted on the Young: Immaturity in Human Development* (New York: Wiley-Blackwell, 2007).

329 'Abdu'l-Bahá, *Paris Talks*, no. 28, p. 103.

330 'Abdu'l-Bahá, *Some Answered Questions*, no. 67, p. 281.

331 'Abdu'l-Bahá. *Tablets of 'Abdu'l-Bahá* (1909; Wilmette, IL: National Spiritual Assembly of the Bahá'ís of the United States, 1980), p. 103.

332 'Abdu'l-Bahá. *'Abdu'l-Bahá in London*, p. 103.

333 M. Bakircioglu, O.P. Carvalho, M. Khurshid et al., 'The essential role of centrosomal NDE1 in human cerebral cortex neurogenesis.', in *The American Journal of Human Genetics*, vol. 88 (2011), issue 5, pp. 523–535.

334 University of California, Los Angeles (UCLA), Health Sciences, 'More sophisticated wiring, not just bigger brain, helped humans evolve beyond chimps, geneticists find', in *ScienceDaily*, 22 August 2012.

335 University of Edinburgh, 'New brain gene gives us edge over apes, study suggests', in *ScienceDaily*, 14 November 2012.

336 Duke University, 'Mouse embryo with big brain: Evolving a bigger brain with human DNA', in *ScienceDaily*, 19 February 2015.

337 K. S. Pollar, S. R. Salama, N. Lambert et al., 'An RNA gene expressed during cortical development evolved rapidly in humans', in *Nature*, no. 443 (14 September 2006), pp. 167–172.

338 'Research finds "unique human DNA" ', *BBC News*, 17 August 2006.

339 'Abdu'l-Bahá, *Some Answered Questions*, no. 49.

340 Ibid.

341 Soil and Water Conservation Society (SWCS), *Soil Biology Primer* (Ankeny, Iowa: Soil and Water Conservation Society, rev. ed. 2000).